Enhanced Oil Recovery Field Case Studies

Enhanced Oil Recovery Field Case Studies

Editor

Bhushan Kulkarni

Enhanced Oil Recovery Field Case Studies

Edited by **Bhushan Kulkarni**

Printed in 2017

ISBN: 978-1-68117-363-4

Library of Congress Control Number: 2015936527

© 2016 by
SCITUS Academics LLC,
616, Corporate Way, Suite 2, 4766,
Valley Cottage, NY 10989

www.scitusacademics.com

Contents

Preface vii

Chapter 1   Electromagnetic Heating of Heavy Oil and Bitumen: A Review of Experimental Studies and Field Applications 1

Albina Mukhametshina and Elena Martynova

Chapter 2   Microbial Enhanced Heavy Oil Recovery by the Aid of Inhabitant Spore-Forming Bacteria: An Insight Review 21

Biji Shibulal, Saif N. Al-Bahry, Yahya M. Al-Wahaibi, Abdulkader E. Elshafie, Ali S. Al-Bemani, and Sanket J. Joshi

Chapter 3   How do Thermal Recovery Methods Affect Wettability Alteration? 53

Abhishek Punase, Amy Zou, and Riza Elputranto

Chapter 4   Laboratory Study on the Potential EOR Use of HPAM/VES Hybrid in High-Temperature and High-Salinity Oil Reservoirs 79

Dingwei Zhu, Jichao Zhang, Yugui Han, Hongyan Wang, and Yujun Feng

Chapter 5   Effect of Polymer Adsorption on Permeability Reduction in Enhanced Oil Recovery 99

Saurabh Mishra, Achinta Bera, and Ajay Mandal

Chapter 6   A Review of CO_2 Sequestration Projects and Application in China 123

Yong Tang, Ruizhi Yang, and Xiaoqiang Bian

Chapter 7   An Experimental and Modeling Study on the Response to Varying Pore Pressure and Reservoir Fluids in the Morrow a Sandstone 153

Aaron V. Wandler, Thomas L. Davis, and Paritosh K. Singh

Chapter 8   **Is There Deep-Seated Subsidence in the Houston-Galveston Area?**...187

Jiangbo Yu, Guoquan Wang, Timothy J. Kearns, and Linqiang Yang

Citations..213
Index...217

Preface

Enhanced oil recovery field case studies bridge the gap between theory and practice in a range of real-world EOR settings. Areas covered include steam and polymer flooding, use of foam, in situ combustion, microorganisms, "smart water"-based EOR in carbonates and sandstones, and many more. Oil industry professionals know that the key to a successful enhanced oil recovery project lies in anticipating the differences between plans and the realities found in the field. This book aids that effort, providing valuable case studies from more than 250 EOR pilot and field applications in a variety of oil fields. The case studies cover practical problems, underlying theoretical and modeling methods, operational parameters, solutions and sensitivity studies, and performance optimization strategies, benefitting academicians and oil company practitioners alike. Strikes an ideal balance between theory and practice.

Editor

Electromagnetic Heating of Heavy Oil and Bitumen: A Review of Experimental Studies and Field Applications

Albina Mukhametshina[1, 2] and Elena Martynova[1, 2]

[1]Harold Vance Department of Petroleum Engineering, Texas A&M University, 3116 TAMU-407 Richardson Building, College Station, TX, USA

[2]Gubkin Russian State University of Oil and Gas, 65 Leninsky Prospekt, Moscow, Russia

ABSTRACT

Viscosity is a major obstacle in the recovery of low API gravity oil resources from heavy oil and bitumen reservoirs. While thermal recovery is usually considered the most effective method for lowering viscosity, for some reservoirs introducing heat with commonly implemented thermal methods is not recommended. For these types of reservoirs, electromagnetic heating is the recommended solution. Electromagnetic heating targets part of the reservoir instead of heating the bulk of the reservoir, which means that the targeted area can be heated up more effectively and with lower heat losses than with other thermal methods. Electromagnetic heating is still relatively new and is not widely used as an alternate or addition to traditional thermal recovery methods. However, studies are being conducted and new technologies proposed that could help increase its use. Therefore, the objective of this study is to investigate the recovery of heavy oil and bitumen reservoirs by electromagnetic heating through the review of existing laboratory studies and field trials.

INTRODUCTION

High-frequency electromagnetic radiation is a relatively new technique for use in enhanced oil recovery methods. It has been tested by theoretic, laboratories and field trial research in Russia [1–10], the United States [11–17], Canada [18–21], and other countries [22–34]. Traditional thermal recovery and well stimulation techniques using hot steam or fluid are not effective in some cases [7, 35] due to prohibitive heat losses from injection wells and reservoirs, low reservoir injectivity (especially for bitumen deposits), steam leakage, large overburden heat loss at thin pay zones, permafrost conditions, and so forth. Furthermore, commonly used thermal recovery methods are not considered environmentally friendly, damaging the hydrogeologic environment and contributing to the greenhouse effect.

The most important thing in electromagnetic heating is that the heat is developed within the material rather than being brought from outside, which means the material is heated more uniformly throughout the medium [27]. Therefore, instead of heating the bulk reservoir volume, part of the reservoir can be targeted and heated more effectively with lower heat loss than other thermal methods. Unlike traditional thermal recovery methods, microwave heating causes friction by vibration of molecules, which results in dielectric heating of the reservoir. Heat and mass transfer in different environments under microwave influence was studied by a number of scientists around the globe, but its application as an EOR method is not yet fully understood. Microwave heating is not used productively because of the lack of reliable information about the processes of heat and mass transfer in a multiphase system in porous media under the influence of electromagnetic radiation, which does not allow effective control. Therefore, current research studies use modeling to discover optimum design parameters for the use of microwave heating in field applications.

REVIEW OF EXPERIMENTAL ELECTROMAGNETIC HEAT STUDIES

The success of near wellbore heating with electromagnetic waves has been proven experimentally [1]. To represent reservoir rock, quartz sand with a 7.7 dielectric constant and a 0.083 tangent was used in the laboratory. A 20% initial water saturation and an 80% initial oil saturation with 16.61 cps (16.61_*10^{-3}Pa·s) gravity and 0.86 g/cm^3 (860 kg/m^3) density oil were maintained. The dielectric constant and the loss tangent of the oil sample were 2.23 and 0.019, respectively. In this study of electric and high-frequency electromagnetic heating of a reservoir model, the temperature of the medium was measured by thermometers located at different points of the experimental setup. In another case, a linear radiator with a length of 87 cm and a diameter of 19 mm was placed in the center of the setup. The linear radiator was connected via acoaxial cable to the generator supplying electromagnetic waves with a frequency of 13.56 MHz. In the experiment, it was discovered that when fluid temperature was exposed to a high-frequency electromagnetic field at the same distance from the radiation source it was greater than one initiated by electric heating. In this case, the thermal conductivity of the medium is affected only slightly. When heat is induced throughout the whole volume of the medium, the amount of heat introduced depends largely on the electrical properties of the medium. The study summarized the advantages of microwave heating over electrical heating as having deeper penetration, quicker heating, and lower heat losses.

Sayakhov [2] discussed the physical foundations of fluid filtration in high-frequency fields. Radial fluid filtration through porous media in a heterogeneous high-frequency electromagnetic field was studied experimentally. A specially designed core holder, filled with a representative porous medium, was located in a coaxial resonator. A high-frequency wave generator was used to generate mw energy, which was directed to the resonator through cables. Studied liquid came into the porous media through the hose first, then through the lumen of the inner conductor of the resonator dripped into a graduated cylinder. Experiments were started with a 200–300 watts power microwave oven at 2400 MHz frequency and 500 watts vibrational power. Temperatures were recorded throughout the experiments via thermocouples inserted in the core holder. Kerosene was used as a representative reservoir fluid. The measurements were performed with and without the influence of a high-frequency electromagnetic field. It was established that exposure to high-frequency electromagnetic fields leads to a sharp increase in flow rate per unit of time and fluid temperature at the outlet. In addition, the flow rate increases dramatically once exposed to the electromagnetic field, while the temperature increases after 10 seconds. After the field is discontinued, a sharp decrease in the flow rate is observed and a gradual cooling of the porous medium takes place.

Experimental studies [3] on the influence of electromagnetic fields with a frequency of $3*10^5$Hz to $6*10^5$ Hz on thermal conductivity of dielectric liquids showed that the thermal conductivity of the liquids increased when exposed. The thermal conductivity increases as the magnitude of the dipole moment of the liquid used in the experiment parameters increases with the frequency and intensity of field.

Fatikhov [5] conducted experimental research on the flow of bitumen oil at different pressure gradients in a high-frequency electromagnetic field. The experiment focused on changes in the volumetric flow rate of filtered oil under different pressures at different temperatures. The initial pressure drop for bitumen from the Mordovo-Karmalskoye deposit in reservoir conditions was 0.003 MPa/m. During the experiment, the pressure drop decreased rapidly as the temperatures were increased, resulting in bituminous oil becoming a Newtonian liquid. Therefore, it was established that the application of electromagnetic heating improves fluid flow behavior and the non-Newtonian properties of bitumen decrease rapidly with increasing temperature.

In 1992, Kasevich et al. [17] studied electromagnetic heating of rock samples at 1 kW (frequency 50.55 MHz) and 200 W (144 MHz) and found that a particular type of rock could be heated to 423 K when exposed to an RF electromagnetic field. The rock, which had low thermal conductivity, heats poorly when hot steam is pumped into the reservoir. They conducted experiments at both normal and formation pressures.

Ovalles et al. [30] used a microwave with 650 watts power to heat core samples saturated with oil of 25 API gravity and 7.7 API gravity (a sample from the Orinoco River Basin). Medium API oil temperatures were measured in 0.5, 1, and 1.5 minute intervals and heavy oil intervals were increased to 1, 5, and 10 minutes. The experimental results were used to test mathematical models and predict the production of three abstract oil reservoirs in Venezuela.

Chakma and Jha [23] conducted laboratory experiments using electromagnetic heating on a scaled thin heavy oil reservoir pay zone model. Gas injection with horizontal wells during electromagnetic heating was achieved. The aim was to decrease oil viscosity with electromagnetic heating and obtain a gas drive with the injected gas. Using nitrogen for the injected gas, they were able to prove that for thin pay zones heating of the wellbore vicinity is sufficient, by achieving oil recoveries as high as 45% of original oil in place compared to estimated primary recovery rates of less than 5%. Recovery achieved by use of the combined method was higher than that of nitrogen injection or electromagnetic heating alone. Chakma and Jha also discussed a number of parameters affecting the results of the combined method, including

- gas injection pressure (when no gas was injected, oil was produced only due to gravity drainage and no significant convective transport occurred; therefore, gas injection provided an oil rate increase with the increasing injection pressure);
- temperature (the initial production rate was not significantly affected by temperature, but later there was an increase in the production rate, meaning that overall recovery increased with temperature);
- electromagnetic frequency (the higher the frequency, the greater the recovery);
- oil viscosity (as expected, a higher oil viscosity leads to a lower recovery for a given electromagnetic frequency, temperature, and gas injection pressure);

- salinity (higher salinity provides higher recovery due to the higher conductivity of saline water compared with distilled water);
- Electrode distance (recovery is similar, but closer electrode spacing provides faster production rates) [23].

Hascakir et al. [24] conducted a laboratory study of microwave-assisted gravity drainage on heavy oil samples from reservoirs in Turkey (Bati Raman, 9.5 API; Garzan, 12 API; and Camurlu, 18 API) using a specially designed novel graphite core holder packed with crushed limestone. Their study described the effects of operational parameters like heating time, waiting period and rock, and fluid properties on the effectiveness of microwave heating. Some of the conclusions made are confirming ones found in Chakma and Jha [23], like the positive effect of high water salinity and water saturation. Hascakir et al. [24] also concluded that water wet conditions are preferable for obtaining higher oil recoveries and that large porosity and permeability are also favorable. When microwave heating is applied to oil samples continuously, higher temperatures are reached, which allows better results to be achieved than with periodic heating when microwave heating is applied for a limited time in periodic intervals. Therefore, higher temperatures allow for better results in continuous heating.

Jha et al. [27] proposed using microwave-assisted gravity drainage (MWAGD) in the Mehsana oil field in India. They heated specially prepared samples with the required characteristics from that field in the laboratory using a microwave with variable power up to 1000 watts operating at 3 GHz frequency, which allowed them to obtain temperature and viscosity profiles of the gravity-drained oil. They described effects of initial oil and water saturations, wettability, porosity, and permeability similar to those found by Hascakir et al. [24] and Chakma and Jha [23].

Jha et al. [27] suggested using MWAGD commercially by drilling one horizontal well and multiple vertical ones with downhole microwave antennas. However, this might not allow deep enough heat penetration, so other options are also proposed such as a combination of two horizontal wells and installing antenna inside the horizontal production well. Vertical separation of the horizontal well pair is approximately 15 meters, which is far more efficient than SAGD in which the separation is around 5 meters. Because crude oil absorbs microwave heat weakly, Jha et al. also proposed increasing thermal conductivity by injecting powdered metallic oxides, chlorides, or activated carbon through a fracture operation. The working principle and description of laboratory applications of such additives to heavy oil can be found in various studies by Hascakir et al., Kershaw et al., and Odenbach [25, 28, 29].

Technical principles of the SAGD method assisted by electromagnetic heating (EM-SAGD process) were reported by Koolman et al. [26]. Inductive heating was initiated in the laboratory using an EM source with a working frequency of 142 kHz. The sample was heated for 10 minutes at a power of 7.2 kW, achieving a rise in the temperature of 7.5 K. Laboratory and field processes were modeled using a numerical simulator, combining electromagnetic and thermal modules. It was specially built and can be applied to field-scale simulations. According to simulation results, a 38% increase in bitumen production was predicted compared to conventional SAGD.

Kovaleva et al. investigated the effects of radio frequency electromagnetic (RF-EM) fields and electrical heating on the mass- and heat-transfer processes in a multicomponent hydrocarbon system flowing in porous media [4,6]. Three different types of experiments were carried out: solvent (kerosene) flooding under the RF-EM field, solvent (kerosene) flooding under electrical heating, and cold solvent (kerosene) flooding. In all three experiments, the physical characteristics of the model and heating conditions (temperatures) were identically maintained. Two series of experiments on models with different granulometric composition of formation were also carried out.

Figure 1 shows the dependence of oil recovery on the volume of solvent injected. This figure demonstrates that the highest oil recovery was obtained by applying an RF-EM field. Kovaleva et al. [4, 6] concluded that following RF-EM influence on the oil-saturated samples the quantity of the received oil is more than the quantity received under electrical (and thermal) processing at identical temperatures of heating of the media. It confirms the additional "nonthermal" action of the electromagnetic field.

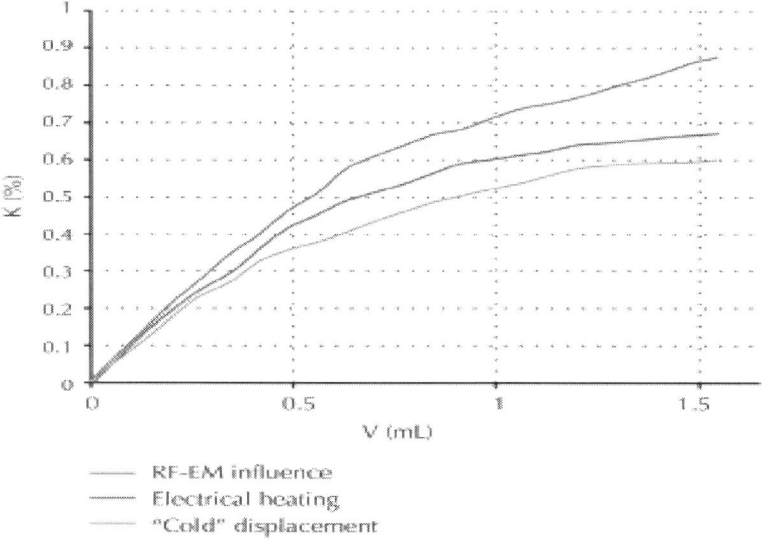

Figure 1: Dependence of oil recovery (K) on the relative volume of solvent injected. Adapted from—[6].

TECHNOLOGIES OF IN SITU ELECTROMAGNETIC HEATING OF HEAVY OIL AND BITUMEN

The first production method applying microwave heating to well production was patented in 1956 [14]. Electromagnetic waves were transferred to the well bottom from the surface through a coaxial system of internal and external pipes (tubing and casing). Interaction of electromagnetic waves with the formation causes the emergence of distributed volumetric heat sources and reduces the viscosity of the reservoir fluid. In 1965, Haagensen [13] described a device for generating high-frequency electromagnetic waves at the mouth of the well and a method of delivering electromagnetic energy through coaxial lines and waveguides to the bottom hole. In 1987, Wilson [16] described a similar device in his work, with some modifications of the radiating element of EM waves.

A huge drawback to the methods described by Haagensen and Wilson [13, 16] is the shallow penetration of electromagnetic waves, and, hence, a low sweep efficiency of heating. When the method described by Ritchey [14] was implemented, there were large losses of electromagnetic energy. Due to the finite conductivity of tubing, they are heated and electromagnetic energy is dissipated in rocks surrounding the well, resulting in large wellbore heat losses, especially if a permafrost layer is present.

Sayakhov et al. [8] proposed a method of recovery that included creating a combustion front with a simultaneous electromagnetic current influence. The environment is heated when exposed to an electromagnetic field, which decreases the viscosity and increases the mobility of crude oil. It is assumed in this method that the EM field continues influencing the reservoir after the combustion is initiated.

Review of Electromagnetic Heating Field Studies

Electromagnetic heating field trials have been carried out in Russia (Bashkortostan and Tatarstan) [2, 9, 10], the United States (California and Utah) [11, 17], and in Canada (Alberta and Saskatchewan) [18–21].

Russia

In Russia, field tests of radio frequency electromagnetic heating of the near-wellbore zone were first launched in 1969 at Well 40/19 in the Ishimbayskoye Oil Field in Bashkortostan and continued in the Yultimirovskoye Bitumen Field in Tatarstan according to Sayakhov et al. [2, 9, 10]. Characteristics of Well 40/19 from the Ishimbayskoye Oil Field are represented in Table 1.

Table 1: Characteristics of Well 40/19 of the Ishimbayskoye Oil Field

Parameter	Value
Depth, m	830
Casing diameter, in.	6

Tubing diameter, in.	2
Flow rate, ton/day	3
Well temperature, K	287–289
Paraffin content, %	2.3
Resins content, %	11
Density, kg/m3	890
Viscosity at 20°C, m2/c	20 * 10–6

Sayakhov et al. [10].

The source of high-frequency electromagnetic energy was a generator providing an optimum oscillation output power of 63 kW at a frequency of 13.56 MHz. Standard RF coaxial cable was used to supply high-frequency electromagnetic energy from the generator to the well. Temperature measurements were carried out using a thermograph at the bottom hole. Temperature was recorded continuously at a fixed depth of 650–655 m (in the open hole, where the radiating element was working) while the generator was operating (Table 2).

Table 2: Dynamics of temperature growth

Heating time, days	Temperature increase at the bottom hole, K
0.5	283
1	290
2	300
3	306
4	311
5	313

Sayakhov et al. [9].

Yultimirovskoye Bitumen Field

RF electromagnetic heating was conducted in the Yultimirovskoye Bitumen Field by Sayakhov in 1980 [2] (Table 3). Two wells spaced 5 meters apart were studied, Well 150 and Well 1.

Table 3: Characteristics of the Yultimirovskoye reservoir

Parameter	Value
Porosity, %	25
Bitumen saturation, %	3.6
Permeability, micm2	0–0.183

Sayakhov [2].

Electromagnetic heating of the bitumen reservoir was conducted in several stages at different conditions. Initially, the RF-EM installation was set to about of 20 kW power. After 36.5 hours, the temperature at the bottom of Well 150 has increased from 282 to 389 K. No temperature change was observed in Well 1. In the next phase, RF-EM installation was reset to approximately 30 kW of power. As a result, the temperature in Well 150 reached 423 K after six hours. It should be noted that the growth rate of temperature increased. In the third phase, the RF-EM installation was set to a maximum of 60 kW, which caused the heating intensity in Well 150 to increase greatly. After 5.5 hours, the temperature in Well 150 increased from 417 to 463 K. Next, the RF-EM unit was turned off for 2 hours, resulting in a temperature drop to 423 K.

During the next 32 hours, the RF-EM installation operated at maximum output, causing the heating of the bottom hole of Well 150 up to 583 K and the bottom hole of Well 1 up to 318 K. In Sayakhov's experiment [2], the fluoroplastic collars that centered the tubing in Well 150 melted (their maximum operating temperature was 573 K). As a result, a short circuit between the casing and tubing occurred and the RF-EM unit broke and was disabled. Before this disruption, the RF-EM installation worked steadily throughout the field experiment.

Deep heat penetration (up to 5 m in the reservoir) was demonstrated by temperature measurements done during the cooling of Well 150's drain zone after the exposure. The temperature decreased from 373 K to 343 K in three days. After a few days of electromagnetic heating, wellbore heat distribution revealed low heat losses to both the overburden and underburden.

The United States

Bakersfield, California, United States

In 1992, Kasevich et al. [17] conducted field tests of RF-EM heating in the United States in the Bakersfield, California, field. The goal was to prove the concept that controlled RF-EM radiation could be used as a thermal EOR method. The production of reservoir fluids was not measured because it was a quality, not quantity, study designed to gain a better understanding of underground processes.

A high-frequency electromagnetic wave generator with a capacity of 25 kW and frequency of 13.56 MHz was used to heat the reservoir at well 100D. Heat penetration was determined by temperature measurements in the surrounding observation wells T10, T20, and T30, which were located 3, 6, and 9 meters from well 100D. Kasevich et al. [17] also proved that the RF producer used could efficiently focus its radiation pattern into the desired region by measuring return loss and electromagnetic radiation. In well T10, situated 3 meters from well 100D, the medium temperature was increased from 293 K to 393 K over 20 hours of EM heating.

Avintaquin Canyon and Asphalt Ridge (Utah)

In 1980, one of the most detailed studies on electromagnetic-heat-based oil recovery was done at the Illinois Institute of Technology Research Institute (IITRI) by Bridges et al. [11]. They carried out extensive research work on the use of the different types of electromagnetic heating for different types of deposits, oil shale, and tar sand.

Bridges et al. [11] tested their IITRI technique of RF electromagnetic heating with two field experiments in Avintaquin Canyon, Utah, USA. Shale that was 6 meters thick was found in outcrops convenient for relatively cheap horizontal experiments. So arrays of holes were drilled and electrodes were inserted to a depth of 1 meter. These tests allowed the researchers to gain experience and to prove it was possible to achieve in situ pyrolysis of oil shale, thus increasing its thermal maturity. The power applied to the formation ranged from 5 kW to 20 kW, with a frequency of 13.56 MHz. As a result of EM heating, temperatures rose to 673 K and 20–30 % of the oil content was collected. However, it should be noted that the amount of produced oil was badly affected by the presence of cracks which allowed light hydrocarbons to escape (evaporate).

In 1981, Bridges et al. [11] conducted field tests on tar sand at Asphalt Ridge, Utah, USA. The first experiment tested the gravity drive bitumen recovery process, designed to prove EM heating concepts and improve equipment design. This experiment used vertical electrode placement and a mined collection chamber and tunnel. It was equipped with a 200 kW radio transmitter and heated 25 m^3 of tar sand. In the first experiment, the roof of the mined chamber was not supported well enough, which resulted in early termination of the experiment. Therefore, the situation had to be fixed by constructing a concrete arch for the second pilot test. The heating power used on tar sand varied from 40 kW to 75 kW with a frequency of 13.56 MHz.

The second test quantified the results of heating over a longer period and at higher temperatures. In this experiment, temperatures exceeded 473 K, and 30 to 35% recovery was achieved in just 20 days. This was encouraging because continuation of heating could have resulted in even higher recovery. The power loss was minimal in all the experiments, which proved the efficiency of heating.

CANADA

The Wildmere Field, Alberta, Canada

According to Spencer [18] commercial EM heating was first introduced in the field at Wildmere, Alberta, Canada (Table 4). The first well was drilled in January 1986 and began producing oil in March of the same year. Before EM heating began in May, the well was producing about 0.95 tonnes/day. After EM heating commenced, production rates increased and soon settled at the level of 3.18 tonnes/day until November 1986, when the well was closed due to technical reasons. Another well in this field increased in production from 1.59 tonnes/day to an average of 4.77 tonnes/day, with the maximum flow rate reaching 9.54 tonnes/day.

Table 4: Characteristics of the reservoir in the Wildmere field, Alberta, Canada

Parameter	Value
Net thickness, m	1
Depth, m	600
Density, kg/m3	987
Oil viscosity at 20°C, Pa·s	20

Spencer [18].

The Lloydminster Heavy Oil Area, Saskatchewan, Canada

In 1988-1989, two electromagnetic stimulation projects were conducted by Davidson [22] in the Lloydminster heavy oil area in Saskatchewan, Canada. Unfortunately the economic potential of the process could not be evaluated from either pilot test, because long-term heating could not be achieved due to equipment failure (casing insulation) and special reservoir conditions. However, the technical results looked promising.

The first pilot well was in Northminster (Saskatchewan, Canada) and produced 11.4 API oil from the Sparky formation. The power was applied to the well in a pulsating manner with a baseline of 20 kW with four-hour spikes of 30 kW (2 daily), in order to reduce the risk of significant damage to the insulation. Later, the power was increased to a 25 kW baseline with 35 kW pulses and finally to a 30 kW baseline with 50 kW peaks. At that point the insulation failed, and the power rates came down to 28 kW. Power rates were later leveled at 47 kW and stayed that way until terminated.

As can be seen from Table 5, water cut and production both reacted positively to electromagnetic stimulation. However, it should be noted that some portion of the increased production is related to the increase in pump speed. Water cut drop can be related directly to the EM effect and improvement in oil mobility. Once the heating has been terminated, technical parameters return rapidly to their initial states.

Table 5: Oil field performance for the North minster pilot: primary and achieved by EM heating

Parameter	Primary	EM heating
Production rate, m3/day	10–12	20
Water cut, %	15–20	10–12
Productivity index, bbl/psi	0.33	0.42
Stimulation ratio		1.27

Davidson [22].

The second pilot well was situated in Lashburn (Saskatchewan, Canada) and produced very viscous 11.4 API crude oil from the Sparky formation. During the reservoir heating phase, the power ranged from 13 to 18 kW. This well has high sand cuts and before electrical power was applied its production was not stable and regularly had to be stimulated by flushing the wellbore. The peak production reached 5.0 m³/day and had begun to decline before electromagnetic heating was applied. Electromagnetic heating reduced the water cut, but the well was still prone to high water cuts after shut-in periods. Oil production also increased when electromagnetic forces were applied, until it reached 9 m³/day. During the initial heating phase of the reservoir, the temperature in the bottom hole increased steadily from 295 K to 309 K, but temperatures began dropping immediately after the power was turned off or when power delivery systems failed.

ESEIEH (Alberta, Canada)

Enhanced Solvent Extraction Incorporating Electromagnetic Heating technology (ESEIEH) has been patented and is currently undergoing tests, as reported by Rassenfoss [20]. The ESEIEH consortium is relying on three oil company partners to help with this testing: Laricina Energy, Nexen, and Suncor Energy. The pilot project is planned to take three years and currently is in the first stage. Field application is expected to start later in 2013. The ESEIEH method combines the familiar horizontal well pair design commonly used in the Canadian oil sands, coupled with heating using RF-EM waves and solvents, such as butane or propane. The company aims to heat the reservoir by running an antenna underground that emits enough energy to raise the temperature to 50°C (120°F).

CONCLUSIONS

A review of electromagnetic heating for enhanced oil recovery was presented in this paper. A number of studies show that electromagnetic heating is a promising method of enhanced oil recovery. However, the studies to date are limited, and only a few field trials have been reported. Most of the current research is based on laboratory experiments or numerical models. It should be noted that this paper did not cover the computer simulations carried out to research the effectiveness of EM heating.

Better understanding of the in situ electromagnetic process is essential and can be achieved by combining laboratory, numerical, and field-scale tests. At the moment it is not possible to assess the efficiency of EM heating or the opportunities for economic applications of it alone or in combination with traditional methods; therefore, more global studies should be conducted.

Even though sustainability of this technology has not yet been completely evaluated, the method definitely should not be overlooked by the industry because of its enormous potential. Attempts should be made to develop viable screening criteria for possible production of heavy oil, oil shale, and tar sand deposits.

REFERENCES

1.   S. Chistyakov, F. Sayakhov, and G. Balabyan, "Experimental study of formations dielectric properties under the influence of high-frequency electromagnetic fields," in University Investigations: Geology and Exploration, pp. 153–156, 1971.

2.   F. Sayakhov, "Particular properties of filtration and fluid flow under the influence of high-frequency electromagnetic field," in Joint University Scientific Book, pp. 108–120, 1980.

3.   B. Savinikh, V. Dyakonov, and A. Usmanov, "The influence of alternating electric currents on the thermal conductivity of dielectric fluids," Journal of Engineering Physics and Thermophysics, no. 2, pp. 269–276, 1981 (Russian).

4.   A. Davletbaev and L. Kovaleva, "Combined RF EM/solvent treatment technique: heavy/extra-heavy oil production model

case study," in Proceedings of the 10th Annual International Conference Petroleum Phase Behavior and Fouling, Rio de Janeiro, Brazil, 2009.

5. M. A. Fatikhov, "Experimental study of bitumen initial pressure gradient in the electromagnetic field,"University Investigations: Oil and Gas, no. 5, pp. 93–94, 1990 (Russian).

6. L. Kovaleva, A. Davletbaev, T. Babadagli, and Z. Stepanova, "Effects of electrical and radio-frequency electromagnetic heating on the mass-transfer process during miscible injection for heavy-oil recovery,"Energy and Fuels, vol. 25, no. 2, pp. 482–486, 2011.

7. G. Malofeev, O. Mirsaetov, and I. Cholovskaya, "Injection of hot fluids for enhanced oil recovery and well stimulation," in Regular and Chaotic Dynamics, Institute of Computer Science, Russialgevsk, Russia, 2008.

8. F. Sayakhov, R. Bulgakov, V. Dyblenko, B. Deshura, and M. Bykov, "About HF heating of bitumen reservoirs," Petroleum Engineering, no. 1, pp. 5–8, 1980 (Russian).

9. F. L. Sayakhov, L. A. Kovaleva, M. A. Fatikhov, and G. A. Khalikov, "Method of thermal effect on oil-bearing formation," SU Patent 1723314, 1992.

10. F. Sayakhov, I. Habibullin, M. Yagudin, and M. Fatyhov, "Technique and technology of thermal well stimulation on the basis electro-thermo-chemical and electromagnetic effects," University Investigations: Oil and Gas, no. 2, pp. 33–42, 1992 (Russian).

11. J. E. Bridges, J. J. Krstansky, A. Taflove, and G. C. Sresty, "The IITRI in situ RF fuel recovery process,"Journal of Microwave Power, vol. 18, no. 1, pp. 3–14, 1983.

12. J. Bridges, "Method for in-situ heat processing of hydrocarbonaceous formation," US Patent 4140180, 1979.

13. A. D. Haagensen, "Oil well microwave tools," Patent USA 3170119, 1965.

14. H. W. Ritchey, "Radiation Heating System, US Patent," Tech. Rep. 2757738, 1956.

15. G. C. Sresty, R. H. Snow, and J. E. Bridges, "Recovery of liquid hydrocarbons from oil shale by electromagnetic heating in-situ," US Patent 4485869, 1984.

16. R. Wilson, "Well production method using microwave heating," US Patent 4485868, 1987.

17. R. S. Kasevich, S. L. Price, D. L. Faust, and M. F. Fontaine, "Pilot testing of a radio frequency heating system for enhanced oil recovery from diatomaceous earth," in Proceedings of the SPE Annual Technical Conference & Exhibition, pp. 105–113, New Orleans, La, USA, September 1994.

18. H. L. Spencer, "Electromagnetic Oil Recovery, Ltd," Calgary, Canada, 1987.

19. F. E. Vermeulen and F. S. Chute, "Electromagnetic techniques in the in-situ recovery of heavy oils,"Journal of Microwave Power, vol. 18, no. 1, pp. 15–29, 1983.

20. S. Rassenfoss, "Seeking more oil, fewer emissions," Journal of Petroleum Technology, vol. 64, no. 9, pp. 34–38, 2012.

21. B. C. W. Mcgee and F. E. Vermeulen, "The mechanisms of electrical heating for the recovery of bitumen from oil sands," Journal of Canadian Petroleum Technology, vol. 46, no. 1, pp. 28–34, 2007.

22. R. J. Davidson, "Electromagnetic stimulation of Lloydminster heavy oil reservoirs: field test results,"Journal of Canadian Petroleum Technology, vol. 34, no. 4, pp. 15–24, 1995.

23. A. Chakma and K. N. Jha, "Heavy-oil recovery from thin pay zones by electromagnetic heating, paper SPE 24817," in Proceedings of the Annual Technical Conference and Exhibition, Society of Petroleum Engineers, Washington, DC, USA, October 1992.

24. B. Hascakir, C. Acar, Schlumberger, B. Demiral, and S. Akin, "Microwave assisted gravity drainage of heavy oils," in Proceedings of the International Petroleum Technology Conference (IPTC '08), pp. 1908–1916, Kuala Lumpur, Malaysia, December 2008.

25. B. Hascakir, T. Babadagli, and S. Akin, "Experimental and numerical modeling of heavy-oil recovery by electrical heating, paper SPE 117669," in Proceedings of the International Thermal Operations and Heavy Oil Symposium (ITOHOS '08), p. 14, Society of Petroleum Engineers, Alberta, Canada, October 2008.

26. M. Koolman, N. Huber, D. Diehl, and B. Wacker, "Electromagnetic heating method to improve steam assisted gravity drainage, paper 1177481," in Proceedings of the International Thermal

Operations and Heavy Oil Symposium (ITOHOS '08), pp. 327–338, Society of Petroleum Engineers, Alberta, Canada, October 2008.

27. K. A. Jha, N. Joshi, and A. Singh, "Applicability and assessment of micro-wave assisted gravity drainage (MWAGD) applications in Mehsana heavy oil field, paper SPE 14591," in Proceedings of the SPE Heavy Oil Conference and Exhibition, Society of Petroleum Engineers, Kuwait City, Kuwait, December 2011.

28. J. R. Kershaw, G. Barrass, and D. Gray, "Chemical nature of coal hydrogenation oils part I. The effect of catalysts," Fuel Processing Technology, vol. 3, no. 2, pp. 115–129, 1980.

29. S. Odenbach, "Ferrofluids—magnetically controlled suspensions," Colloids and Surfaces A, vol. 217, no. 1–3, pp. 171–178, 2003.

30. C. Ovalles, A. Fonseca, A. Lara et al., "Opportunities of downhole dielectric heating in Venezuela: three case studies involving medium, heavy and extra-heavy crude oil reservoirs, paper SPE 78980," inProceedings of the International Thermal Operations and Heavy Oil Symposium and International Horizontal Well Technology Conference, Alberta, Canada, November 2002.

31. M. A. Ayrapetyan, "About oil fields development prospects by high-frequency currents electrical fields," in Materials of KSSR Institute of Oil, pp. 38–52, 1958.

32. M. A. Ayrapetyan, V. S. Velikanov, and E. Ya. Magnikov, "Reservoir high-frequency heating investigations," in Materials of KSSR Institute of Oil, pp. 113–124, 1959.

33. M. A. Carrizales, L. W. Lake, and R. T. Johns, "Production improvement of heavy-oil recovery by using electromagnetic heating, paper SPE 115723," in Proceedings of the SPE Annual Technical Conference and Exhibition (ATCE '08), Denver, Colo, USA, September 2008.

34. A. D. Hiebert, F. E. Vermeulen, F. S. Chute, and C. E. Capjack, "Numerical simulation results for the electrical heating of Athabasca oil-sand formations," SPE Reservoir Engineering, vol. 1, no. 1, pp. 76–84, 1986.

35. J. Burge, P. Surio, and M. Combarnu, Thermal Methods of Enhanced Oil Recovery, Nedra Publishing, Moscow, Russia, 1988.

Chapter 2

Microbial Enhanced Heavy Oil Recovery by the Aid of Inhabitant Spore-Forming Bacteria: An Insight Review

Biji Shibulal[1], Saif N. Al-Bahry[1], Yahya M. Al-Wahaibi[2], Abdulkader E. Elshafie[1], Ali S. Al-Bemani[2], and Sanket J. Joshi[1,3]

[1]Department of Biology, College of Science, Sultan Qaboos University, 123 Muscat, Oman

[2]Petroleum and Chemical Engineering Department, College of Engineering, Sultan Qaboos University, 123 Muscat, Oman

[3]Central Analytical and Applied Research Unit, College of Science, Sultan Qaboos University, 123 Muscat, Oman

ABSTRACT

Crude oil is the major source of energy worldwide being exploited as a source of economy, including Oman. As the price of crude oil increases and crude oil reserves collapse, exploitation of oil resources in mature reservoirs is essential for meeting future energy demands. As conventional recovery methods currently used have become less efficient for the needs, there is a continuous demand of developing a new technology which helps in the upgradation of heavy crude oil. Microbial enhanced oil recovery (MEOR) is an important tertiary oil recovery method which is cost-effective and eco-friendly technology to drive the residual oil trapped in the reservoirs. The potential of microorganisms to degrade heavy crude oil to reduce viscosity is considered to be very effective in MEOR. Earlier studies of MEOR (1950s) were based on three broad areas: injection, dispersion, and propagation of microorganisms in petroleum reservoirs; selective degradation of oil components to improve flow characteristics; and production of metabolites by microorganisms and their effects. Since thermophilic spore-forming bacteria can thrive in very extreme conditions in oil reservoirs, they are the most suitable organisms for the purpose. This paper contains the review of work done with thermophilic spore-forming bacteria by different researchers.

BACKGROUND

Oil productions have been experiencing decline in many parts of the world due to the oil field maturity, and example of such includes the major oil fields in the North Sea [1]. Another major factor which causes downgrade is the increasing energy demands due to global population growth and the difficulty in discovering new oil fields as an alternative to the exploited oil fields. Therefore, there is an urge to find out alternative technologies to increase oil recovery from existing oil fields around the world. It is a fact that fossil fuels will still remain the key source of energy, regardless of the gross investments in other energy sources such as biofuels, solar energy, and wind energy. Current global energy production from fossil fuels represents about 80–90% with oil and gas typifying about 60% [2]. Cossé [3] stated that during the process of oil production, between 30 and 40% of oil can be contributed by primary oil recovery, while additional 15–25% can be recovered by secondary methods such as water injection leaving behind about 35–55% of oil as residual oil in the reservoirs. The focus of many enhanced oil recovery technologies is this residual oil, and it amounts to about 2–4 trillion barrels [4] or about 67% of the total oil reserves [5]. For many oil companies, residual oil recovery is at present unavoidable, and so there is a perpetual hunt for a cheap and efficient technology which will raise the global oil production as well as the productive life of many oil fields. The recovery of this residual oil is accomplished by enhanced oil recovery (EOR) or tertiary recovery methods which are used in oil industry to increase the production of crude oil. Most common tertiary recovery methods include chemical flooding, miscible CO_2 injection, and thermally enhanced oil recovery method which uses heat as a main source for the additional oil recovery [6]. Large quantities of residual oil in the depleted oil reservoirs could be regained by these EOR methods as the current primary and secondary extraction methods leave about two-thirds of the original oil in the reservoir. One of the potential EOR methods is microbial enhanced oil recovery (MEOR), which employs microorganisms to pull out the remaining oil from the reservoirs. Up to 50% of the residual oil can be extracted by this exceptionally low operating cost technology [7, 8]. The field trials of MEOR method project a chance to reverse the declining trend of oil production or at least to maintain a curve with a positive slope. This is achieved by the alteration of chemical and physical properties of reservoir rocks and crude oil by the microbial growth and metabolites produced [9]. MEOR can overcome the main hindrances of efficient oil recovery such as low reservoir permeability, high viscosity of the crude oil, and high oil-water interfacial tensions, which in turn result in high capillary forces retaining the oil within the reservoir rock [10].

THE REASONS FOR OIL TO GET LEFT BEHIND

The fundamental cause for leaving oil behind is economics. In general, the process of recovering oil from any conventional reservoir requires (a) a pathway which connects oil in the pore space of a reservoir to the surface and (b) sufficient energy in the reservoir to drive the oil to the surface. Lack of these requirements in the environment results in oil getting left behind. In this case, it is not economical to implement incremental development activities. In addition, all of the theoretically displaceable oil cannot be recovered, even if there is a pathway and adequate reservoir energy, due to the physics of fluid displacement in porous media [11].

ENHANCED OIL RECOVERY

The residual crude oil in reservoirs is up to 67% of the total petroleum reserves in the world, which in turn represents the relative inefficiency of the primary and secondary production techniques. Extraction of this trapped oil can be achieved by injecting chemicals (polymers or surfactants), gases (carbon dioxide, hydrocarbons, or nitrogen), or steam into the reservoir. The chemicals used for EOR must be compatible with the physical and chemical environments of oil reservoirs. The varying permeability of petroleum reservoirs is also a major concern in EOR processes. When water is injected to displace the oil, it preferentially flows through areas of highest permeability and bypasses much of the oil [12]. Thus, the conventional EOR methods to recover the entrapped crude oil seem not to be very efficient.

MICROBIAL ENHANCED OIL RECOVERY (MEOR)

MEOR is a tertiary oil recovery technique. Recovering oil usually requires three stages. At the primary recovery only 12% to 15% of the oil in the well is recovered without the need to introduce other substances into the well. The oil well is then flooded with water or other substances to drive out an additional oil (15% to 20%) from the well which is known as the secondary recovery. Tertiary recovery is the last phase which is accomplished through several different methods, including MEOR, for the additional extraction of trapped oil from the well. In principle, the process of MEOR results in some beneficial effects such as formation of stable oil-water emulsions reduced interfacial tension and clogging the high permeable zones. In in situ MEOR method, bacteria inoculated with water in to the well will progress into high-permeability zones at first. Then at a later stage they will grow and occlude those zones due to their size and the negative charge on their cell surface. This scenario helps to increase the sweep efficiency, and thus a more efficient oil recovery can be achieved [11, 13].

Microorganisms can synthesize useful products by fermenting low-cost substrates or raw materials. Therefore, MEOR can substitute chemical enhanced oil recovery (CEOR), which is a very pricey technology. In MEOR, the chosen microbial strains are used to synthesize compounds analogous to those used in CEOR processes which are very expensive, to increase the recovery of oil from depleted and marginal reservoirs. Furthermore, microbial products are biodegradable and have low toxicity [7, 14, 15]. Microbial technologies are becoming approved universally as lucrative and eco-friendly approaches to improve oil production [16, 17].

MEOR OUTCOMES

MEOR is based on two absolute justifications. Oil advancement through porous media is expedited by modifying the interfacial properties of the oil-water minerals. In such a system, microbial activity alters fluidity (viscosity reduction, miscible flooding); displacement efficiency (decrease of interfacial tension, increase of permeability); sweep efficiency (mobility control, selective plugging); and driving force (reservoir pressure).

The second principle is known as upgrading. In this case, the degradation of heavy oils into lighter ones occurs by microbial activity. Instead, it can also aid in the removal of sulphur from heavy oils as well as the removal of heavy metals.

Continuous research and successful applications affirm the fact that MEOR can be viewed as a potent technology [8, 22, 23] despite the existing disagreement by some groups [24]. However, successful MEOR field applications reported are specific for each well and published information to support economic advantages is lacking. MEOR is, therefore, considered as one of the promising future research areas with great preference as identified by the Oil and Gas in the 21st Century Task Force [24]. This is probably because MEOR is an alternate technology that may help in recovering the 377 billion barrels of oil that are unrecoverable by conventional technologies [8].

THE BYGONE DAYS OF MEOR

It was Beckman in 1926 [25] who suggested for the first time that microbes could be used to recover oil from porous media. Between 1926 and 1940, not many studies were held on this topic. In the 1940s, Zobell [26] started a series of systematic laboratory findings which marked the beginning of a new era of petroleum microbiology research with application in oil recovery. According to Zobell the main mechanisms behind oil release from porous media are processes such as bacterial metabolites that break up inorganic carbonates; bacterial gases which reduce the viscosity of oil, thereby increasing its flow; surface-active substances or wetting agents produced by some bacteria; and the high affinity of bacteria for solids to crowd off the oil films, processes by which bacterial products (gases, acids, solvents, surface-active agents, and cell biomass) releasing oil from the sand pack columns in wet labs were patented by Zobell. Later Updegraff et al. repeated [27, 28] Zobell's experiments and patented [29] the process which is based on the bacterial byproducts produced from cheap substrates like molasses to assist the oil recovery. The first field test was carried out in the Lisbon field, Union County, AR [30]. Kuznetsov et al. [31] concluded that anaerobic bacteria present in the oil deposits can utilize oil to form gaseous products (CH_4, H_2, CO_2, N_2). Kuznetsov's work demonstrated the technology of microbial flora activation of reservoirs, later advanced by Ivanov et al. [32]. Extensive research on MEOR was conducted in the 1960s and 1970s, in Czechoslovakia, Hungary, and Poland [33–35]. The field trials were based on the injection of mixed anaerobic or facultative anaerobic bacteria (Clostridium, Bacillus, Pseudomonas, Arthrobacterium, Micrococcus, Peptococcus, Mycobacterium, etc.) selected on their ability to generate gases, acids, solvents, polymers, surfactants, and cell biomass. At the same time, another technology named as selective plugging recovery has been recognized as an important additional mechanism for improving the oil recovery from water floods. This is achieved by producing polysaccharide slime in situ by an injected microbial system based on molasses. Microbes producing biopolymers of xanthan or scleroglucan types as viscosifying agents were isolated, which greatly enhanced oil recovery [36–38]. The investigations during 1970–2000 have demonstrated the basic nature and existence of indigenous microbiota in oil reservoirs, as well as reservoir characteristics essential to a successful MEOR application. It was also proved that the cyclic microbial recovery (single well stimulation), microbial flooding recovery, and selective plugging recovery are very effective. The technology based on activation of stratal microbiota was successfully developed in former Soviet Union [32, 39]. It can be concluded that the petroleum crisis during 1970s led to substantial MEOR research and later became a scientifically identified EOR method, supported by research projects carried out all over the world in countries such as the USA, Canada, Australia, China, Russia, Romania, Poland, Hungary, Czech Republic, Great Britain, Germany, Norway, and Bulgaria. Many international meetings were periodically organized on the MEOR topic with the publication of proceedings carrying the advances in the knowledge and practice of MEOR techniques. It is important to recognize and acknowledge the role of the U.S. Department of Energy (DOE), which sponsored MEOR basic research and field trials, as well as periodically organizing international meetings. Several books on MEOR were also published [40–42]. Grula et al. [43] developed a microbial screening method to isolate an anaerobic Clostridium species that produced gases, acids, alcohols, and surfactants. But all those strains isolated showed intolerance to high salt concentrations (>7%) which remained as a major problem. Success of in situ MEOR processes depends upon isolating microorganisms that can survive and produce the desired metabolic products in reservoirs containing hydrocarbons and saline water. Continuous investigations were done on different microbial species such asClostridium species, Bacillus species, and Enterobacter for better adaptation to reservoir conditions. By the end of the 1990s, MEOR was recognized as a scientific and interdisciplinary technique for the increase of oil recovery.

In 1995, a survey of 322 MEOR projects in the USA showed that 81% of the projects successfully increased oil production, and neither of them had shown reduced oil production [7]. Today, MEOR technologies are well suited for application, when there is a need for oil crisis at a rate of 3 to 4%/year. Since 1980, the abolition of stripper wells has increased to 175% [9], and accordingly, within 15–25 years, the USA could have access to less than 25% of its remaining oil resources. MEOR technologies were very slowly recognized by industry even though a long history of MEOR activity exists, due to the lack of published data especially in widely available journals, as well as too little cooperation between microbiologists, reservoirs engineers, geologists, economists, and owner operators.

LABORATORY AND FIELD MEOR PROJECTS

Zobell [44] patented a process for the secondary oil recovery, using anaerobic, hydrocarbon-utilizing, and sulfate-reducing bacteria such as Desulfovibrio species in situ. He reported that the oil recovery mechanism was similar to Clostridium, where bacterial cells (and the hydrogenase enzyme system) produces the acids and ammonium hydroxide by using CO_2, water and nitrates present in the reservoir, which helps to enhance the release of oil from reservoir rock when supplied with nutrients.

Various "agroindustrial carbohydrates based" substrates are proposed as a suitable "carbon source" for MEOR applications, like molasses [17, 45]. Updegraff and Wren [27] proposed that fermentative bacteria such asDesulfovibrio use nutrients such as molasses to produce large amounts of organic acids and carbon dioxide to enhance oil recovery in wet labs. The process was patented by them in spite of the major drawback ofDesulfovibrio species producing hydrogen sulfide which is not suitable for MEOR processes. MEOR research team at Sultan Qaboos University, Oman, have reported isolation, identification, and bioproducts production by spore-forming Bacilli spp., and its potential role in enhancing oil recovery at laboratory scale [13, 17–21]. Bond [46] injected 5,000 gal of agar medium containing sand and Desulfovibrio hydrocarbonoclasticus, which is no longer a valid species into a sandstone reservoir at a depth of 3,000 ft. The well initially produced 15 bbl/day. After the inoculum injection, the well was shut in for 3 months for the bacterial growth and action. The well, when it started the production again, produced 25 bbl/day.

Hitzman [47] patented a process of injecting bacterial spores along with nutrients into a reservoir. The spores would germinate in the reservoir and enhance oil recovery from reservoir rock. A medium containing molasses and spores of Clostridium roseum was passed through the sand-packed column saturated with oil and showed about 30% increase in release of oil.

Patents by Hitzman [36, 47] used microorganisms that utilized injected polymers and the byproducts of CO_2floods, to produce products such as gases, acids, solvents, and surfactants for EOR.

In polymer floods, the injected organisms feed on polymer that is adsorbed on the reservoir rock. In CO_2 floods, the microbes feed on soluble compounds of carbon, nitrogen, and sulfur left behind by the CO_2-crude oil slug. The process was demonstrated in sand-pack, but no core or field tests are reported.

Knapp et al. [48] reported the isolation of 22 microorganisms that produce biopolymers and emulsifiers. Among them, one strain could thrive at 10% salt concentrations, over a pH range of 4.6 to 9.0, at temperatures up to 50°C, in presence of crude oil. They demonstrated that glucose, ammonium sulfate, and potassium phosphate were easily transported through sandstone cores. Viable bacterial cells in aqueous solutions of 2% NaCl and 0.01% $CaC1_2$ injected into these cores were not recovered in the effluent. The cores were inoculated with bacteria and nutrients such as glucose were added which resulted in a significant decrease in permeability. This could be because of the plugging of pores by the bacterial mass. The prominent bacteria indigenous to all of the cores treated were found as Pseudomonas sp., Bacillus sp., and Actinomycetes. A major problem in these experiments was the determination of the amount of plugging caused by injected bacteria and the amount by inhabitant ones. The problem existed even when cores were steam-sterilized and autoclaved. "Sterilization" of cores with chlorine dioxide helped to get rid of the problem, but the bacterial populations returned after 48 h incubation.

Johnson [49] studied 150 stripper wells in the USA that produced, on an average, 2 bbl/day, with no well head pressure. The reservoir porosities were 10 to 30%, depths 200 to 1,000 ft, with an average reservoir temperature of 38°C. In his study, he inoculated a mixed culture of Bacillus and Clostridium spp. (1 to 10 gal) with crude molasses and mineral salts as nutrients. Approximately 10 to 14 days were needed for the optimal growth of cells in the treated area of the reservoir. The results varied, but an average of 20 to 30% additional oil-in-place was recovered.

The preliminary field tests done by Petrogen, Inc., during 1977–88 in 24 wells with varying depths from 300 to 4,600 ft., demonstrated a pressure increase of 10 to 200 psi in 75% of the wells. Four wells doubled production for 6 months, and 12 increased production by 50% for 3 months. The average production increase was indicated as 42%; however, the final results remain to be reported [50]. Jack et al. [51] considered that emulsification of viscous crude oil in situ is not a feasible method for EOR since transporting the bacteria through the reservoir rock would face some difficulties. Yarbrough and Coty [30] reported a field test performed by them in 1954 in Arkansas, in which Clostridium acetobutylicum was injected along with a 2% solution of beet molasses in fresh water during a 6-month period. 70 days after starting the injection,freshwater breakthrough occurred at the production well. Fermentation products such as short-chain fatty acids, CO_2, and traces of ethanol, 1-butanol, and acetone and sugars were found 80 to 90 days after the injection started. There was no increase in hydrogen content. Production of oil increased from 0.6 bbl/day to 2.1 bbl/day. Field test studies were not conducted.

The first MEOR project in the Rocky Mountains was started in 1983 [52]. An independent oil operator acquired three field service operations from Petroleum Bio-Resources Company. These were (a) a reservoir field conditioning system to avoid plugging; (b) use of a microorganism that produces gas and surfactant; and (c) use of a microorganism that produces a polysaccharide for mobility control. It was stated that production immediately doubled due to well stimulation and also increased oil recovery from 26 to 60 bbl/day, probably due to mobilization of oil by microorganisms, and water flooding was also noticed in other fields [53]. Bryant and Douglas [54] demonstrated the oil recovery efficiency of several different bacterial strains in Berea sandstone cores. They reported that additional 32% oil was recovered as compared to water flooding, and some spore-forming bacteria even showed 50–60% additional oil recovery. Berea sandstone core experiments showed that selected microbial strains could recover up to 72% of the heavy oil (API 14° and 17°) left after water flooding.

Field Tests

Kuznetsov [55] reported that bacteria were present in certain oil-gas-bearing strata in the Saratov and Buguruslan areas of the USSR in such numbers that large quantities of CO_2 were generated (depth was approximately 3,300 ft). Certainly methane was also formed. In the later works, Kuznetsov et al. [31] introduced a mixed culture of aerobic and anaerobic bacteria with acid-hydrolyzed substances from peat and soils and shut in the well for 6 months, and after that the well was opened for production [50]. The rate of oil production rose from 275 to 300 bbl/day; however, 4 months later it had fallen to 270 bbl/day. Field tests were done by Dostálek and Spurny [33] in Czechoslovakia where they injected sulfate-reducing (Desulfovibrio) and hydrocarbon-utilizing (Pseudomonas) bacteria with nutrients (molasses). During six-month experiment period, the daily average oil production increased by nearly 7%. No further work has been reported since 1958. Heningen et al. [56] reported on two field tests performed in the Netherlands, in which they used Betacoccus dextranicus in a sucrose-molasses medium of 10% total sugar content and obtained a 30% increase in cumulative oil recovery. A mixed culture of slime-forming bacteria in 50% molasses was used in the subsequent field trial. The oil-to-water production ratio changed to 1 : 20 compared to 1 : 50 before the treatment.

In Hungary, to recover naphthenic crude, Jaranyi et al. [57] utilized a mixture of anaerobic thermophilic bacteria that fermented molasses. They also tried with raw sewage as an inoculum (100 L, along with 20 to 40,000 kg molasses) in their later trials (1969-70). The deepest reservoir was 8,200 ft, where the pressure was 228 atm and the temperature was 97°C. In 70% of the reservoirs tested, the introduced microbial populations showed positive results on overall oil recovery.

Karaskiewicz [58] conducted 18 field trials in Poland between 1961 and 1969. Microbial cultures were obtained from soil and water samples which were collected from the nearby areas of the oil fields and from sugar factory waters. The mixed culture includes the genera Arthrobacter, Clostridium, Mycobacterium, Peptococcus, andPseudomonas grown in 10 L bottles with formation water plus 4% molasses, incubated at 32°C. The wells ranged in depth from 1,650 to 5,000 ft. The rate of additional oil recovery ranged from 20 to 200% of the original production rate. An additional supply of nutrients was proved to be a major factor for the increased oil recovery. Lazar [59] published an extensive review of MEOR work done in Romania during the last decade, in which he discussed three major areas in MEOR including (a) isolation of the bacterial population from the formation water of the reservoir; (b) adaptation of these microorganisms in wet lab for oil release; and (c) field testing of such adapted cultures. Seven wells were treated with microbial formulations, and he concluded that the bacterial population caused an increase of oil flow up to 200% for 1 to 5 years in 2 out of 7 reservoirs (the other five were unaffected), and much information about the ecology of the reservoir is needed before initiating any MEOR activity. A list of various reported successful MEOR applications at laboratory scale and field are listed in Table 1.

Table 1: Successful laboratory and field MEOR applications [7, 13, 17–21]

Country	Biological systems used
USA	Pure or mixed cultures of Bacillus, Clostridium, Pseudomonas, and Gram-negative rods; mixed cultures of hydrocarbon degrading bacteria; mixed cultures of marine source bacteria; spore suspension of Clostridium; indigenous stratal microflora; slime-forming bacteria; ultramicrobacteria
Russia	Pure cultures of C. tyrobutiricum; bacteria mixed cultures; indigenous microflora of water injection and water formation; activated sludge bacteria; naturally occurring microbiota of industrial (food) wastes
China	Mixed enriched bacterial cultures of Bacillus, Bacteroides, Eubacterium, Fusobacterium, Pseudomonas; slime-forming bacteria:Brevibacterium viscogenes, Corynebacterium gumiform, Xanthomonas campestris
Australia	Ultramicrobacteria with surface active properties
Bulgaria	Indigenous oil-oxidizing bacteria from water injection and water formation
Canada	Pure culture of Leuconostoc mesenteroides
Former Czechoslovakia	Hydrocarbon oxidizing bacteria (predominant Pseudomonas sp.); sulfate-reducing bacteria
England	Naturally occurring anaerobic strain, high generator of acids; special starved bacteria, good producers of exopolymers
Former East Germany	Mixed cultures of thermophilic Bacillus and Clostridium from indigenous brine microflora
Hungary	Mixed sewage-sludge bacteria cultures (predominant: Clostridium, Desulfovibrio, Pseudomonas)
Norway	Nitrate-reducing bacteria naturally occurring in North Sea water

Oman	Autochthonous spore-forming bacteria from oil wells and oil contaminated soil
Poland	Mixed bacteria cultures (Arthrobacter, Clostridium, Mycobacterium, Peptococcus, Pseudomonas)
Romania	Adapted mixed enrichment cultures (predominant: Bacillus, Clostridium, Pseudomonas, and other Gram-negative rods)
Saudi Arabia	Adequate bacterial inoculum according to requirements of each technology
The Netherlands	Slime-forming bacteria (Betacoccus dextranicus)
Trinidad-Tobago	Facultative anaerobic bacteria high producers of gases
Venezuela	Adapted mixed enrichment cultures

HEAVY OIL

Heavy crude oil or extra heavy crude oil is a type of crude oil which does not flow easily. It is referred to as "heavy" because of its density or specific gravity, which is higher than that of light crude oil. Heavy crude oil has been defined as any liquid petroleum with API gravity less than 20°, which means its specific gravity is greater than 0.933. This type of oil forms due to the exposure of crude oil to bacteria [60]. Production, transportation, and refining of heavy crude oil are much difficult compared to light crude oil. The largest reserves of heavy oil in the world are located in the north of the Orinoco River in Venezuela (Energy Information Administration, 2001) the same amount as the conventional oil reserves of Saudi Arabia, but 30 or more countries are known to have such heavy crude oil reserves. Heavy crude oil is closely related to oil sands; the main difference is that oil sands generally do not flow at all. Canada has large reserves of oil sands, located north and northeast of Edmonton, Alberta. Physical properties that distinguish heavy crudes from lighter ones include higher viscosity and specific gravity, as well as heavier molecular composition. Extra heavy oil from the Orinoco region has a viscosity of over 10,000 centipoise and 10° API gravity. A diluent is added at regular distances in pipeline carrying heavy oil to increase the flow rate [61].

Field Tests

Heavy crude oil plays a major role in the economics of petroleum development. The heavy oil resources in the world are more than twice those of conventional light crude oil. In October 2009, the USGS updated the Orinoco tar sands (Venezuela) recoverable value to 513 billion barrels (8.16 × 1010 m³) (USGS. 11 January 2010), making this area the world's first recoverable oil deposit, ahead of Saudi Arabia and Canada [61]. The price of heavy crude oil slashes as compared to light oil due to increased refining costs and high sulphur content. The high viscosity and density also make production more difficult. On the other hand, large quantities of heavy crudes have been discovered in the Americas including Canada, Venezuela, and California. Another reason can be the relatively shallow depth of heavy oil fields (often less than 3000 feet) which contributes to lower production costs [62]. Special techniques are being developed for exploration and production of heavy oil.

Chemical Properties

Heavy oil contains asphaltenes and resins. It is "heavy" (dense and viscous) due to the high ratio of aromatics and naphthenes to paraffins (linear alkanes) and high amounts of NSOs (nitrogen, sulfur, oxygen, and heavy metals). The carbon chain in heavy oil has over 60 carbon atoms which results in a high boiling point and molecular weight. For example, the viscosity of Venezuela's Orinoco extra-heavy crude oil lies in the range of 1000–5000 cP (1–5 Pa·s), while Canadian extra-heavy crude has a viscosity in the range of 5000–10,000 cP (5–10 Pa·s), about the same as molasses, and higher (up to 100,000 cP or 100 Pa·s for the most viscous commercially exploitable deposits) [62]. A definition from the Chevron Phillips Chemical Company is as follows.

The "heaviness" of heavy oil is primarily the result of a relatively high proportion of a mixed bag of complex, high molecular weight, nonparaffinic compounds, and a low proportion of volatile, low molecular weight compounds. Heavy oils typically contain very little paraffin and may or may not contain high levels of asphaltenes.

DEVELOPMENT OF HEAVY OIL RESERVES IN OMAN

The first oil discovery in the Sultanate of Oman was accomplished in 1956, when City Services Company drilled Marmul-1 well. But the discovery was not considered as a commercial discovery because the oil found was heavy compared to oil discoveries in the Middle East at that time. In 1962, Petroleum Development of Oman (PDO) exploration activities ended up in achieving commercial discovery of oil in Yibal field, followed by discoveries in Natih and Fahud fields in 1963 and 1964, respectively. These discoveries marked the birth of Oman as an oil producing country. The result of these successes in discovering and production of oil inspired the Government to sign two new agreements to explore oil and gas in 1973 and another two in 1975 with other international oil companies. By 2009, the number of active oil fields reached 135. Over these days, Oman has been continuously applying efforts to improve the recovery of its oil reserves and has adopted EOR techniques on a large scale. These initiatives helped the Sultanate to increase their oil production capability to nearly 1 million barrels per day (bpd) from 714300 bpd averaged in 2008 [63]. This has also changed the outlook for its oil industry which is now estimated to have at least 40 years of life ahead of it [64]. Al-Ghubar South's discovery in 2009 was the most auspicious discovery for Oman. According to the Ministry of Oil and Gas, this discovery could add as much as 1 billion barrels to reserves. Two other convincing discoveries, including that in Malaan West and Taliah in the Lekhwair cluster in northwest Oman, were made which will stretch the baseline production in the future [65].

First Trials

The first export of Omani oil took place on July 27,1967. In the beginning, oil production increased steadily to 341000 barrels per day in 1975 and in 1984; the average daily production reached around 400000 barrels per day. Petroleum Development Oman (PDO)—the largest oilfield operator in Oman—started a series of EOR trials in 1986 due to low recovery of oil because of the complex geology of the reservoirs. The trials proved successful and Oman slowly started implementing EOR thereby boosting the production to a current level of nearly 900000 bpd. EOR projects result in 5–15 percent increment in reserves and PDO expects its EOR projects to contribute around 35 percent of its total production by 2020. So Oman is considered as a country which is pushing the limits of EOR technology [66].

Miscible Gas Injection

Miscible gas injection involves pumping gas to oil wells. These gases that are being used for this purpose are often toxic which will dissolve in the oil and eventually lead to higher flow rates. This technique is currently at its operations in the Harweel oil field cluster [65].

Steam Injection

Qarn Alam is the world's first full field EOR project and also the largest of its kind in the world. Thermally assisted gas oil gravity drainage (TAGOGD), a sophisticated method, is employed due to the characteristics of the fractured carbonate reservoir, as the oil is highly viscous and a very low percentage of recovery is feasible by conventional oil extraction method.

Polymer Injection

Marmul field is located in south Oman. It is characterized by heavy viscous crude that is difficult to extract by traditional recovery methods. The reservoir has a viscosity of around 90 cP. The reservoir's sweep efficiency was modified by viscosifying the water with the addition of polyacrylamide polymers and then injected in the reservoir through polymer injection wells. The polymer flooding at Marmul field will increase a further recovery by 8000 bpd. By this technique, 10–15% increase in recovery levels from the Marmul reservoirs is predicted.

EOR Projects in Oman Oil Fields

Miscible gas injection has been applied in Harweel oil field which resulted in an additional production of 40,000 bbl/day. Thermal EOR methods are being deployed at Mukhaizna, Marmul, Amal-East,

Amal-West, and Qarn Alam fields. Mukhaizna has already increased production to 50,000 bbl/day, and the other fields, Amal-East and Amal-West, are expected to raise the production to 23,000 bbl/day by 2018. Furthermore, the steam injection at Qarn Alam is supposed to enhance the production by 40,000 bbl/day by 2015. This is achieved by a novel process in which steam drains oil to lower producer wells. At projects such as Marmul, with its heavy oil reserves, injecting polymer fluid has seen to be more effective.

Other EOR projects include Karim cluster, a cluster of 18 oil fields flowing to the Nimr production facility, in which PDO is aiming to boost up the production. In Harweel cluster, PDO estimates approximately 40 percentage increase in the next five years. Also with Rima clusters, using EOR techniques, much gain is expected (US Energy Information Administration, 2012).

ROLE OF MICROBES IN BIODEGRADATION OF HEAVY CRUDE OIL

Degradation of oil is one of the most important parts of the MEOR by which the oil's viscosity and freezing point are reduced which in turn will increase the oil's flow in situ. Heavy oils are rich in gum and asphaltene, having characteristics such as freezing point, low flow ability, difficult oil recovery, and high recovery cost [67].

Microbe can improve the physical characteristics of heavy oil in two ways: (1) by degrading heavy oil fractions, thereby decreasing the average molecular weight of heavy oil; and (2) the byproducts of microbial metabolism, such as biological surface active substance, acid, and gas, which can reduce the viscosity of oil considerably. Gum and asphaltene present in heavy oil have high molecular weight and polarity; meanwhile they are one of the main factors making the oil recovery difficult [68]. Usually, microbe hardly degrades them [69]. Zhang et al. [70] separated a variety of microbes from environments rich in petroleum and done a series of experiments using mixed microbial consortia, which can effectively degrade heavy oil, even gum and asphaltene; these microbes act by lowering the viscosity and freezing point of heavy oil and thereby improving the physical and chemical characters of heavy oils.

In some cases, using microbial consortia with different properties (ability to degrade heavy oil fractions and biosurfactant production) thereby applying different mechanisms might have a desired effect for enhanced oil recovery [71]. There are a lot of microbes having the ability to degrade hydrocarbons by using them as carbon sources [72]. Interesting results for the microbial n-alkane degradation have been reported during the past decades [73–76]. Extensive studies have been made on strains of Gordonia amicalis which have shown to be a potent degrader of large n-alkanes under aerobic and anaerobic conditions [77]; many Pseudomonas species have the ability to degrade lighter hydrocarbons with carbon chain length C_{12}–C_{32}, and heavier hydrocarbons with carbon chain length of C_{36}–C_{40} [78, 79]; and a thermophilic Bacillus strain that degrades only long-chain (C_{15}–C_{36}) hydrocarbons but not short-chain (C_{8}–C_{14}) n-alkanes [80] has also been reported. The ability of biosurfactant-producing indigenous Bacillus strains to degrade the higher fractions of crude oil and aid in the enhancement of its flow characteristics has also been studied for a petroleum reservoir in the Daqing Oilfield [81]. The MEOR team in the Sultan Qaboos University, Oman, found that a consortia of Bacillus strains form oil contaminated soil degraded heavy chain oil (C_{50}–C_{70}) to (C_{11}–C_{20}). Many microorganisms contain genes coding for the enzymes responsible for degrading petroleum hydrocarbons. Some microorganisms degrade alkanes (normal, branched, and cyclic paraffins), others aromatics, and others both paraffinic and aromatic hydrocarbons [82–84]. The most readily degraded alkanes are considered to be in in the range of C_{10} to C_{26}, but low-molecular-weight aromatics, such as benzene, toluene, and xylene, which are considered as the toxic compounds found in petroleum, are also readily biodegraded by many marine microorganisms. As the complexity of the structures (those with branches and/or condensed ring structures) increases, it will be more resistant for biodegradation, which means only fewer microorganisms can degrade those structures and the biodegradation rates would be much lower than the rates for the simpler hydrocarbon structures found in petroleum. The higher the number of methyl-branched components or condensed aromatic rings, the slower the rates of biodegradation and the greater the probability of accumulating partially oxidized intermediary metabolites.

Petroleum contains numerous compounds of varying structural complexities. The residual mixture formed after petroleum biodegradation may resist further biodegradation. Crude oils are never completely degraded and always result in some complex residue which appears as a black tar containing a high proportion of asphaltic compounds. The toxicity and bioavailability of the residual mixture are very low as long as it does not coat and suffocate an area, thus becoming an inert environmental contaminant with no toxic effects on environment [60].

About 10% of the total bacterial population in hydrocarbon-contaminated marine environments is hydrocarbon-degrading bacterial populations [82]. The major metabolic pathways for hydrocarbon biodegradation have been elucidated [85]. The initial steps in the biodegradation of hydrocarbons by bacteria are the oxidation of the oil by oxygenases. Alkanes are subsequently converted to carboxylic acids that are further biodegraded via -oxidation (the central metabolic pathway for the utilization of fatty acids from lipids, which results in the formation of acetate, enters into the tricarboxylic acid cycle). Generally aromatic hydrocarbon rings are hydroxylated to form diols, which are then eventually cleaved to form catechols which are subsequently degraded to intermediates of the tricarboxylic acid cycle. Interestingly, the intermediates resulting from bacterial action are with differing stereochemistry usually cis-diols, which are biologically inactive. With bacteria being the dominant hydrocarbon degraders in the marine environment, the products of aromatic hydrocarbons biodegradation will detoxify them and do not produce potential carcinogens. The complete biodegradation (mineralization) of hydrocarbons produces environmentally safe end products such as carbon dioxide and water, as well as cell biomass (largely protein) which will eventually enter into the food web.

MICROBIAL CANDIDATES INVOLVED IN CRUDE OIL DEGRADATION

Thermophilic Spore-Forming Bacteria Involved in Biodegradation of Heavy Crude Oil for MEOR

Many varieties of microbes are identified and isolated from different petroleum reservoirs which comprise several ecological niches, including sulfate reducers [86–88], sulfur reducers [86], methanogen [88], fermentative bacteria [89, 90], manganese and iron reducers [91], and dibenzothiophene-degrading bacteria [92]. Although many bacteria are isolated from many reservoirs, those which can be applied to MEOR are fewer.

Many researchers have been engaged in studying thermophiles. It is reported that 140 species of 70 genera of thermophiles have been discovered from high temperature environments with wide applications [93]. In the Shengli oil field of East China, where extreme physical conditions exist with temperature ranging 60–90°C and depth of 1000–2000 m, most of the reservoirs are under EOR. This harsh environment seems to be unsuitable for microbial growth. But some thermophiles have been isolated which helps in EOR [86].

There are many kinds of Bacillus, which are distributed widely, but those which have application on crude oil recovery are very few [94, 95]. B. subtilis and B. licheniformis strains have been repeatedly isolated from many oil reservoirs as well as oil contaminated samples, thus confirming the adaptability of these species [17–19,95–101]. The properties of B. subtilis have been reported in much literature [17–19, 96, 102–104], but the isolation and its action on crude oil have been scarcely reported [95].

It is recognized that the thermophiles possess enzymes which are more resistant to physical and chemical denaturation. Their faster growth rates also serve as another major advantage. Relative studies suggest that thermophilic hydrocarbon degraders of Bacillus, Thermus, Thermococcus, and Thermotoga species occurring in natural high-temperature or sulfur-rich environments are of special significance [105]. Wang et al. [106] isolated functional bacteria from high temperature petroleum reservoirs. Three thermophilic hydrocarbon-degrading bacteria, which belonged to Bacillus sp., Geobacillus sp., and Petrobacter sp., could tolerate 55°C in obligate anaerobic condition. These strains could utilize crude oil as carbon source with the degradation rate of 56.5%, 70.01%, and 31.78%, respectively, along with the viscosity reduction rate of 40%, 54.55%, and 29.09%, meanwhile the solidify points of crude oil were reduced by 3.7, 5.2, and 3.1°C.

Hao et al. [107] isolated a hydrocarbon-degrading bacterium, strain SB-1, from oil-contaminated soil samples collected at the Shengli oil field in east China. Based on 16S rDNA sequence, the strain was identified as B. subtilis. The bacteria degraded 39.33% of crude oil, 57.01% of the saturated fractions, 25.63% of the resins, and 12.15% of the aromatic fractions within 12 days. In addition, more than 50% of the alkanes were removed by the strain; the highest degradation rate was shown as 81.03% for C_{36}–C_{40}, and the lowest degradation rate being 51.47% for C_{31}–C_{35}. The results of this study concluded that B. subtilis SB-1 is a potent strain in degrading oil pollutants in soil.

Sanchez et al. [108] isolated thermophilic bacteria enriched from the formation waters of a Venezuelan oil field. The reservoir, located at Maracaibo Lake, has a temperature of 60–80°C and a pressure of 1,200–1,500 psi. The main fermentative byproducts were alcohols, short chain fatty acids, and gases when grown in media with industrial wastes as carbon source.

A strain of B. stearothermophilus (Geobacillus) was isolated from oil-contaminated Kuwaiti desert capable of growing on C_{15}–C_{17} [109], and two strains of G. jurassicus were isolated from a high temperature petroleum reservoir capable of growing on C_6–C_{16} [110]. B. thermoleovorans strain isolated from deep subterranean petroleum reservoirs was shown to degrade n-alkane up to C_{23} at 70°C [111]. Thermophilic, glucose-fermenting, strictly anaerobic, rod-shaped bacterium, Thermotoga hypogea sp. strain SEBR 6459T (T = type strain), was isolated from an African oil-producing well [112] and T. elfii strain SEBR 6459 by Ravot et. al. [113]. Al-Bahry et al. [18–21, 96] reported 33 genera and 58 species identified from Omani oil wells. All of the identified microbial genera were first reported in Oman, with Caminicella sporogenes for the first time reported from oil fields. Most of the identified microorganisms were found to be anaerobic, thermophilic, and halophilic and produced biogases, biosolvents, and biosurfactants as by-products, which may be potentially applicable in MEOR.

Various bioremediation and biodegradation agents are commercially available consisting of microbial cultures or microbial enzymes or both. The US Environmental Protection Agency National Contingency Plan released a product schedule report on August 2013 [114]. Also various laboratory screening reports are available for these commercial products [115].

CONCLUSIONS

Given the scarcity of the literature on thermophilic spore-forming bacteria involved in MEOR for crude oil biodegradation, there is a clear need for further laboratory research. While significant progress has been made, we still need to rigorously examine this mechanism of MEOR.

REFERENCES

1.   K. Aleklett, M. Höök, K. Jakobsson, M. Lardelli, S. Snowden, and B. Söderbergh, "The peak of the oil age—analyzing the world oil production reference scenario in world energy outlook 2008," Energy Policy, vol. 38, no. 3, pp. 1398–1414, 2010.

2.   W. Graus, M. Roglieri, P. Jaworski, L. Alberio, and E. Worrell, "The promise of carbon capture and storage: evaluating the capture-readiness of new EU fossil fuel power plants," Climate Policy, vol. 11, no. 1, pp. 789–812, 2011.

3.   R. Cossé, Basics of Reservoir Engineering, Pure and Applied Geophysics, Éditions Technip, 1993.

4.   C. Hall, P. Tharakan, J. Hallock, C. Cleveland, and M. Jefferson, "Hydrocarbons and the evolution of human culture," Nature, vol. 426, no. 6964, pp. 318–322, 2003.

5.   R. S. Bryant, A. K. Stepp, K. M. Bertus, T. E. Burchfield, and M. Dennis, "Microbial-enhanced waterflooding field pilots," Developments in Petroleum Science, vol. 39, pp. 289–306, 1993.

6.   L. W. Lake, Enhanced Oil Recovery, Prentice Hall, Englewood Cliffs, NJ, USA, 1989.

7.   I. Lazar, I. G. Petrisor, and T. F. Yen, "Microbial enhanced oil recovery (MEOR)," Petroleum Science and Technology, vol. 25, no. 11, pp. 1353–1366, 2007.

8.   R. Sen, "Biotechnology in petroleum recovery: the microbial EOR," Progress in Energy and Combustion Science, vol. 34, no. 6, pp. 714–724, 2008.

9.   D. O. Hitzman, "Microbial enhanced oil recovery—the time is now," Developments in Petroleum Science, vol. 31, pp. 11–20, 1991.

10. B. Bubela, "A comparison of strategies for enhanced oil recovery using in situ and ex situ produced biosurfactants," Surfactant Science Series, vol. 25, pp. 143–161, 1987.

11. G. S. Derek, "Microbiological methods for the enhancement of oil recovery," Biotechnology and Genetic Engineering Reviews, vol. 1, no. 1, pp. 187–222, 1984.

12. S. Rebeka, "Potential uses of microorganisms in petroleum recovery technology," in Proceedings of the Oklahoma Academy of Science, 1987.

13. R. Al-Hattali, H. Al-Sulaimani, Y. Al-Wahaibi et al., "Microbial biomass for improving sweep efficiency in fractured carbonate reservoir using date molasses as renewable feed substrate," in Proceedings of the SPE Annual Technical Conference and Exhibition, San Antonio, Tex, USA, 2012.

14. H. Suthar, K. Hingurao, A. Desai, and A. Nerurkar, "Evaluation of bioemulsifier mediated microbial enhanced oil recovery using sand pack column," Journal of Microbiological Methods, vol. 75, no. 2, pp. 225–230, 2008.

15. I. M. Banat, A. Franzetti, I. Gandolfi et al., "Microbial biosurfactants production, applications and future potential," Applied Microbiology and Biotechnology, vol. 87, no. 2, pp. 427–444, 2010.

16. A. K. Sarkar, J. C. Goursaud, M. M. Sharma, and G. Georgiou, "Critical evaluation of MEOR processes,"In Situ, vol. 13, no. 4, pp. 207–238, 1989.

17. S. N. Al-Bahry, Y. M. Al-Wahaibi, A. E. Elshafie et al., "Biosurfactant production by Bacillus subtilis B20 using date molasses and its possible application in enhanced oil recovery," International Biodeterioration and Biodegradation, vol. 81, pp. 141–146, 2013.

18. H. Al-Sulaimani, Y. Al-Wahaibi, S. N. Al-Bahry et al., "Experimental investigation of biosurfactants produced by Bacillus species and their potential for MEOR in Omani oil field," in Proceedings of the SPE EOR Conference at Oil and Gas West Asia 2010 (OGWA ‹10), pp. 378–386, Muscat, Oman, April 2010.

19. H. Al-Sulaimani, Y. Al-Wahaibi, S. Al-Bahry et al., "Optimization and partial characterization of biosurfactants produced by

Bacillus species and their potential for ex-situ enhanced oil recovery," SPE Journal, vol. 16, no. 3, pp. 672–682, 2011.

20.   H. Al-Sulaimani, Y. Al-Wahaibi, S. N. Al-Bahry et al., "Residual-oil recovery through injection of biosurfactant, chemical surfactant, and mixtures of both under reservoir temperatures: induced-wettability and interfacial-tension effects," SPE Reservoir Evaluation and Engineering, vol. 15, no. 2, pp. 210–217, 2012.

21.   Y. Al-Wahaibi, H. Al-Hadrami, S. Al-Bahry, A. Elshafie, A. Al-Bemani, and S. Joshi, "Residual oil recovery via injection of biosurfactant and chemical surfactant following hot water injection in Middle East heavy oil field," in Proceeding of the SPE Heavy Oil Conference, Alberta, Canada, June 2013.

22.   K. Fujiwara, Y. Sugai, N. Yazawa, K. Ohno, C. X. Hong, and H. Enomoto, "Biotechnological approach for development of microbial enhanced oil recovery technique," Studies in Surface Science and Catalysis, vol. 151, pp. 405–445, 2004.

23.   H. Al-Sulaimani, S. Joshi, Y. Al-Wahaibi, S. N. Al-Bahry, A. Elshafie, and A. Al-Bemani, "Microbial biotechnology for enhancing oil recovery: current developments and future prospects," Biotechnology, Bioinformatics and Bioengineering Journal, vol. 1, no. 2, pp. 147–158, 2011.

24.   A. R. Awan, R. Teigland, and J. Kleppe, "A survey of North Sea enhanced-oil-recovery projects initiated during the years 1975 to 2005," SPE Reservoir Evaluation and Engineering, vol. 11, no. 3, pp. 497–512, 2008.

25.   J. W. Beckman, "The action of bacteria on mineral oil," Industrial and Engineering Chemistry, News Edition, vol. 4, pp. 23–26, 1926.

26.   C. E. Zobell, "Bacterial release of oil from oil-bearing materials," World Oil, vol. 126, pp. 36–47, 1947.

27.   D. M. Updegraff and G. B. Wren, "The release of oil from petroleum-bearing materials by sulfate-reducing bacteria," Applied Microbiology, vol. 2, no. 6, pp. 309–322, 1954.

28.   J. B. Davis and D. M. Updegraff, "Microbiology in the petroleum industry," Bacteriological Reviews, vol. 18, no. 4, pp. 215–238, 1954.

29. D. M. Updegraff, "Recovery of petroleum oil," US Patent No. 2.807.570, 1957.

30. H. F. Yarbrough and V. F. Coty, "Microbial enhanced oil recovery from the upper crustaceous nacatoch formation," in Proceedings of the International Conference on Microbial Enhancement of Oil Recovery, 1983.

31. S. I. Kuznetsov, M. V. Ivanov, and N. N. Lyalikowa, Introduction to Geological Microbiology, McGraw-Hill, New York, NY, USA, 1963.

32. M. V. Ivanov, S. S. Belyaev, M. A. Zyakun, A. V. Bondar, and S. K. Laurinavichus, "Microbiological formation of methane in the oil field development," Moscova, vol. 11, 1983.

33. M. Dostálek and M. Spurny, "Bacterial release of oil. A preliminary trial in an oil deposit," Folia Biologica, vol. 4, pp. 166–172, 1958.

34. M. Dienes and I. Yaranyi, "Increase of oil recovery by introducing anaerobic bacteria into the formation Demjen field," Hungary Koolaj as Fodgas, vol. 106, no. 7, pp. 205–208, 1973.

35. I. Karaskiewicz, "The application of microbiological method for secondary oil recovery from the Carpathian crude oil reservoir," Widawnistwo "SLASK", pp. 1–67, 1974.

36. D. O. Hitzman, "Review of microbial enhanced oil recovery field tests," in Proceedings of the Applications of Microorganisms to Petroleum Technology, U.S. Department of Energy, 1988.

37. I. Lazar, "MEOR field trials carried out over the world during the past 35 years," in Microbial Enhancement of Oil Recovery— Recent Advances, E. C. Donaldson, Ed., 1991.

38. I. Lazar, "International MEOR applications for marginal wells," Pakistan Journal of Hydrocarbon Research, vol. 10, pp. 11–30, 1998.

39. M. V. Ivanov, S. S. Belyaev, I. A. Borzenkov, I. F. Glumov, and P. B. Ibatulin, "Additional oil production during field trials in Russia," in Microbial Enhancement of Oil Recovery—Recent Advances, E. Premuzic and A. Woodhead, Eds., 1993.

40. J. E. Zajic, D. G. Cooper, T. R. Jack, and N. Kosaric, Microbial Enhanced Oil Recovery, Penn Well Books, Tulsa, Okla, USA, 1983.

41. T. F. Yen, State of the Art Review on Microbial Enhanced Oil Recovery, NSF OIR-8405134, Los Angeles, Calif, USA, 1986.

42. E. C. Donaldson, G. V. Chilingarian, and T. F. Yen, Microbial Enhanced Oil Recovery, Elsevier, New York, NY, USA, 1989.

43. E. A. Grula, H. H. Russell, D. Bryant, M. Kanaga, and M. Hart, "Isolation and screening of Clostridia for possible use in microbially enhanced oil recovery," in Proceedings of the Microbial Enhanced Oil Recovery, Afton, Okla, USA, 1982.

44. C. E. Zobell, "Bacteriological process for treatment of fluid-bearing earth formations," US Patent No. 2, 413, 278, 1946.

45. S. Joshi, C. Bharucha, S. Jha, S. Yadav, A. Nerurkar, and A. J. Desai, "Biosurfactant production using molasses and whey under thermophilic conditions," Bioresource Technology, vol. 99, no. 1, pp. 195–199, 2008.

46. D. C. Bond, Bacteriological Method of Oil Recovery, Pure Oil Company, Rock Hill, SC, USA, 1961.

47. D. O. Hitzman, "Microbiological secondary recovery of oil," U.S. Patent 3, 032, 472, 1962.

48. R. M. Knapp, M. J. McInerney, D. E. Menzie, and G. E. Jenneman, "The use of microorganisms in enhanced oil recovery," in First Annual Report to the Department of Energy, 1983.

49. A. C. Johnson, "Microbial oil release technique for enhanced oil recovery," in Proceedings of the Conference on Microbiological Processes Useful in Enhanced Oil Recovery, San Diego, Calif, USA, 1979.

50. D. O. Hitzman, "Petroleum microbiology and the history of its role in enhanced oil recovery," inProceedings of the International Conference on Microbial Enhancement of Oil Recovery, U. S. Department of Energy, Bartlesville, Okla, USA, 1982.

51. T. R. Jack, B. G. Thompson, and E. D. Blasio, "The potential for use of microbes in the production of heavy oil," in Proceedings of the International Conference on Microbial Enhanced Oil Recovery, Afton, Okla, USA, 1982.

52. R. Rountree, "Rocky mountain oil history," Western Oil Reporter, vol. 4, p. 77, 1984.

53. J. E. Zajic, Proceedings of the 1st International MEOR Workshop, U.S. Department of Energy Report No. DOE/BC/10852-1, 1986.

54. R. S. Bryant and J. Douglas, "Evaluation of Microbial Systems in Porous Media for EOR," SPE Reservoir Engineering, vol. 3, no. 2, pp. 489–495, 1988.

55. S. I. Kuznetsov, "Possibilities of production of methane in oil fields of Saratov and Buguruslan,"Mikrobiologiia, vol. 19, no. 3, pp. 193–202, 1950.

56. J. V. Heningen, A. J. DeHann, and J. D. Jansen, "Process for the recovery of petroleum from rocks," Netherlands Patent 80, 580, 1958.

57. I. Jaranyi, L. Kiss, C. Salanczy, and J. Szolnoki, "Alteration of some characteristics of oil-wells through the effects of microbial treatment," in Proceedings of the 3rd International Science Conference on Geochemistry, 1963.

58. J. Karaskiewicz, "Studies on increasing petroleum oil recovery from Carpathian deposits using bacteria,"Nafta (Petroleum), vol. 21, pp. 144–149, 1975.

59. I. Lazar, "Microbially enhanced oil recovery in Romania," in Proceedings of the International Conference on Microbial Enhanced Oil Recovery, Afton, Okla, USA, 1982.

60. M. B. Dusseault, "Comparing Venezuelan and Canadian heavy oil and tar sands," in Proceedings of the Petroleum Society›s Canadian International Petroleum Conference, Calgary, Canada.

61. H. A. Rodriguez, P. Vaca, O. Gonzalez, and M. C. de Mirabal, "Integrated study of a heavy oil reservoir in the Orinoco Belt: a field case simulation," in Proceedings of the SPE Reservoir Simulation Symposium, pp. 309–310, June 1997.

62. S. Chopra and L. Lines, "Introduction to this special section: heavy oil," The Leading Edge, vol. 27, no. 9, pp. 1104–1106, 2008.

63. http://www.engineerlive.com/content/23933.

64. http://www.mog.gov.om/english/AboutUs/TheHistoryofOilGas/tabid/117/Default.aspx.

65. http://www.eia.gov/countries/cab.cfm?fips=MU.

66. http://www.reportlinker.com/p0155667/oman-oil-and-Gas-Report-Q4.html.

67. G. L. Lei, "Research and application of microbial enhanced oil recovery," Acta Petrolei Sinica, vol. 22, no. 2, pp. 56–61, 2001.

68. Y. S. Peng, H. S. Ji, and C. X. Liang, Field Research of Microbial Enhanced Oil Recovery, Petroleum Industry Press, Beijing, China, 1997.

69. T. S. Zhang, G. Z. Lan, L. Deng, and X. G. Deng, "Experiments on heavy oil degradation and enhancing oil recovery by microbial treatments," Acta Petrolei Sinica, vol. 22, no. 1, pp. 54–57, 2001.

70. T. S. Zhang, X. Chen, G. Z. Lan, and Z. Jiang, "Microbial degradation influences on heavy oil characters and MEOR test," in Proceedings of the 18th World Petroleum Congress, Johannesburg, South Africa, September 2005.

71. L. Jinfeng, M. Lijun, M. Bozhong, L. Rulin, N. Fangtian, and Z. Jiaxi, "The field pilot of microbial enhanced oil recovery in a high temperature petroleum reservoir," Journal of Petroleum Science and Engineering, vol. 48, no. 3-4, pp. 265–271, 2005.

72. A. Wentzel, T. E. Ellingsen, H. K. Kotlar, S. B. Zotchev, and M. Throne-Holst, "Bacterial metabolism of long-chain n-alkanes," Applied Microbiology and Biotechnology, vol. 76, no. 6, pp. 1209–1221, 2007.

73. V. G. Grishchenkov, R. T. Townsend, T. J. McDonald, R. L. Autenrieth, J. S. Bonner, and A. M. Boronin, "Degradation of petroleum hydrocarbons by facultative anaerobic bacteria under aerobic and anaerobic conditions," Process Biochemistry, vol. 35, no. 9, pp. 889–896, 2000.

74. J. S. Sabirova, M. Ferrer, D. Regenhardt, K. N. Timmis, and P. N. Golyshin, "Proteomic insights into metabolic adaptations in Alcanivorax borkumensis induced by alkane utilization," Journal of Bacteriology, vol. 188, no. 11, pp. 3763–3773, 2006.

75. A. Etoumi, I. El Musrati, B. El Gammoudi, and M. El Behlil, "The reduction of wax precipitation in waxy crude oils by Pseudomonas species," Journal of Industrial Microbiology and Biotechnology, vol. 35, no. 11, pp. 1241–1245, 2008.

76. M. Binazadeh, I. A. Karimi, and Z. Li, "Fast biodegradation of long chain n-alkanes and crude oil at high concentrations with Rhodococcus sp. Moj-3449," Enzyme and Microbial Technology, vol. 45, no. 3, pp. 195–202, 2009.

77. D. H. Hao, J. Q. Lin, X. Song, J. Lin, Y. J. Su, and Y. B. Qu, "Isolation, identification, and performance studies of a novel paraffin-

degrading bacterium of Gordonia amicalis LH3," Biotechnology and Bioprocess Engineering, vol. 13, no. 1, pp. 61–68, 2008.

78. I. M. Banat, "Biosurfactants production and possible uses in microbial enhanced oil recovery and oil pollution remediation: a review," Bioresource Technology, vol. 51, no. 1, pp. 1–12, 1995.

79. M. Hasanuzzaman, A. Ueno, H. Ito et al., "Degradation of long-chain n-alkanes (C36 and C40) byPseudomonas aeruginosa strain WatG," International Biodeterioration and Biodegradation, vol. 59, no. 1, pp. 40–43, 2007.

80. L. Wang, Y. Tang, S. Wang et al., "Isolation and characterization of a novel thermophilic Bacillus strain degrading long-chain n-alkanes," Extremophiles, vol. 10, no. 4, pp. 347–356, 2006.

81. Y. H. She, F. Zhang, J. J. Xia et al., "Investigation of biosurfactant-producing indigenous microorganisms that enhance residue oil recovery in an oil reservoir after polymer flooding," Applied Biochemistry and Biotechnology, vol. 163, no. 2, pp. 223–234, 2011.

82. R. M. Atlas, "Microbial degradation of petroleum hydrocarbons: an environmental perspective,"Microbiological Reviews, vol. 45, no. 1, pp. 180–209, 1981.

83. J. G. Leahy and R. R. Colwell, "Microbial degradation of hydrocarbons in the environment,"Microbiological Reviews, vol. 54, no. 3, pp. 305–315, 1990.

84. R. M. Atlas and R. Bartha, "Hydrocarbon biodegradation and oil spill bioremediation," Advances in Microbial Ecology, vol. 12, pp. 287–338, 1992.

85. R. M. Atlas, Petroleum Microbiology, Macmillan, New York, NY, USA, 1984.

86. K. O. Stetter, R. Huber, E. Blöchl et al., "Hyperthermophilic archaea are thriving in deep North Sea and Alaskan oil reservoirs," Nature, vol. 365, no. 6448, pp. 743–745, 1993.

87. C. Tardy-Jacquenod, P. Caumette, R. Matheron, C. Lanau, O. Arnauld, and M. Magot, "Characterization of sulfate-reducing bacteria isolated from oil-field waters," Canadian Journal of Microbiology, vol. 42, no. 3, pp. 259–266, 1996.

88. T. K. Ng, P. J. Weimer, and L. J. Gawel, "Possible nonanthropogenic origin of two methanogenic isolates from oil producing

wells in the San Miguelito field, Ventura County, California," Geomicrobiology Journal, vol. 7, no. 3, pp. 185–192, 1989.

89. M. E. Davey, W. A. Wood, R. Key, K. Nakamura, and D. A. Stahl, "Isolation of three species of Geotoga and Petrotoga: two new genera, representing a new lineage in the bacterial line of descent distantly related to the 'Thermotogales'," Systematic and Applied Microbiology, vol. 16, no. 2, pp. 191–200, 1993.

90. G. S. Grassia, K. M. McLean, P. Glénat, J. Bauld, and A. J. Sheehy, "A systematic survey for thermophilic fermentative bacteria and archaea in high temperature petroleum reservoirs," FEMS Microbiology Ecology, vol. 21, no. 1, pp. 47–58, 1996.

91. A. C. Greene, B. K. C. Patel, and A. J. Sheehy, "Deferribacter thermophilus gen. nov., sp. nov., a novel thermophilic manganese- and iron-reducing bacterium isolated from a petroleum reservoir,"International Journal of Systematic Bacteriology, vol. 47, no. 2, pp. 505–509, 1997.

92. A. Bahrami, S. A. Shojaosadati, and G. Mohebali, "Biodegradation of dibenzothiophene by thermophilic bacteria," Biotechnology Letters, vol. 23, no. 11, pp. 899–901, 2001.

93. Z. He, Q. Peng, and J. Chen, Biology of Thermophiles, Scientific Press, Beijing, China, 2000.

94. J. Li, B. Lian, J. Hao, J. Zhao, and L. Zhu, "Non-parallelism between the effect of microbial flocculants on sewerage disposal and the flocculation rate," Chinese Journal of Geochemistry, vol. 25, no. 2, pp. 139–142, 2006.

95. G. E. Jenneman, M. J. McInerney, R. M. Knapp et al., "A halotolerant, biosurfactant producing Bacillusspecies potentially useful for enhanced oil recovery," Developments in Industrial Microbiology, vol. 24, pp. 485–492, 1983.

96. S. N. Al-Bahry, A. Elshafie, Y. Al-Wahaibi et al., "Microbial consortia in Oman oil fields: a possible use in enhanced oil recovery," Journal of Microbiologyand Biotechnology, vol. 23, no. 1, pp. 106–117, 2013.

97. M. M. Yakimov, K. N. Timmis, V. Wray, and H. L. Fredrickson, "Characterization of a new lipopeptide surfactant produced by thermotolerant and halotolerant subsurface Bacillus licheniformis BAS50,"Applied and Environmental Microbiology, vol. 61, no. 5, pp. 1706–1713, 1995.

98.　S. M. M. Dastgheib, M. A. Amoozegar, E. Elahi, S. Asad, and I. M. Banat, "Bioemulsifier production by a halothermophilic Bacillus strain with potential applications in microbially enhanced oil recovery,"Biotechnology Letters, vol. 30, no. 2, pp. 263–270, 2008.

99.　H. Ghojavand, F. Vahabzadeh, M. Mehranian et al., "Isolation of thermotolerant, halotolerant, facultative biosurfactant-producing bacteria," Applied Microbiology and Biotechnology, vol. 80, no. 6, pp. 1073–1085, 2008.

100. N. Youssef, M. S. Elshahed, and M. J. McInerney, "Microbial processes in oil fields. Culprits, problems, and opportunities," Advances in Applied Microbiology, vol. 66, pp. 141–251, 2009.

101. D. R. Simpson, N. R. Natraj, M. J. McInerney, and K. E. Duncan, "Biosurfactant-producing Bacillus are present in produced brines from Oklahoma oil reservoirs with a wide range of salinities," Applied Microbiology and Biotechnology, vol. 91, no. 4, pp. 1083–1093, 2011.

102. D. G. Cooper, C. R. Macdonald, S. J. B. Duff, and N. Kosaric, "Enhanced production of surfactin fromBacillus subtilis by continuous product removal and metal cation additions," Applied and Environmental Microbiology, vol. 42, no. 3, pp. 408–412, 1981.

103. C. N. Mulligan, T. Y. K. Chow, and B. F. Gibbs, "Enhanced biosurfactant production by a mutantBacillus subtilis strain," Applied Microbiology and Biotechnology, vol. 31, no. 5-6, pp. 486–489, 1989.

104. G. Zheng and M. F. Slavik, "Isolation, partial purification and characterization of a bacteriocin produced by a newly isolated Bacillus subtilis strain," Letters in Applied Microbiology, vol. 28, no. 5, pp. 363–367, 1999.

105. H. Feitkenhauer, R. Müller, and H. Märkl, "Degradation of polycyclic aromatic hydrocarbons and long chain alkanes at 60–70°C by Thermus and Bacillus spp.," Biodegradation, vol. 14, no. 6, pp. 367–372, 2003.

106. J. Wang, T. Ma, J. Liu et al., "Isolation of functional bacteria guided by PCR-DGGE technology from high temperature petroleum reservoirs," Huan Jing Ke Xue, vol. 29, no. 2, pp. 462–468, 2008.

107. R. Hao, M. Lv, and A. Lu, "Biodegradation of crude oil in soil by Bacillus subtilis SB-1," Current Topics in Biotechnology, vol. 6, pp. 49–55, 2011.

108. G. Sanchez, A. Marin, L. Vierma, and T. P. Eugene, Isolation of Thermophilic Bacteria from a Venezuelan Oil Field, Developments in Petroleum Science, Elsevier, New York, NY, USA, 1993.

109. N. A. Sorkhoh, A. S. Ibrahim, M. A. Ghannoum, and S. S. Radwan, "High-temperature hydrocarbon degradation by Bacillus stearothermophilus from oil-polluted Kuwaiti desert," Applied Microbiology and Biotechnology, vol. 39, no. 1, pp. 123–126, 1993.

110. T. N. Nazina, D. S. Sokolova, A. A. Grigoryan et al., "Geobacillus jurassicus sp. nov., a new thermophilic bacterium isolated from a high-temperature petroleum reservoir, and the validation of the Geobacillusspecies," Systematic and Applied Microbiology, vol. 28, no. 1, pp. 43–53, 2005.

111. T. Kato, M. Haruki, T. Imanaka, M. Morikawa, and S. Kanaya, "Isolation and characterization of psychrotrophic bacteria from oil-reservoir water and oil sands," Applied Microbiology and Biotechnology, vol. 55, no. 6, pp. 794–800, 2001.

112. M. L. Fardeau, B. Ollivier, B. K. C. Patel et al., "Thermotoga hypogea sp. nov., a Xylanolytic, thermophilic bacterium from an oil-producing well," International Journal of Systematic Bacteriology, vol. 47, no. 4, pp. 1013–1019, 1997.

113. G. Ravot, M. Magot, M. L. Fardeau et al., "Thermotoga elfii sp. nov., a novel thermophilic bacterium from an African oil-producing well," International Journal of Systematic Bacteriology, vol. 45, no. 2, pp. 308–314, 1995.

114. U.S. Environmental Protection Agency, National Contingency Plan, Product Schedule, 2013.

115. R. J. Portier and L. M. Basirico, Laboratory Screening of Commercial Bioremediation Agents for the Deepwater Horizon Spill Response, Department of Environmental Sciences, Louisiana State University, Baton Rouge, La, USA, 2011.

Chapter 3

How do Thermal Recovery Methods Affect Wettability Alteration?

Abhishek Punase, Amy Zou, and Riza Elputranto

Department of Petroleum Engineering, Texas A&M University, College Station, TX 77840, USA

ABSTRACT

We will investigate the effect of temperature on wettability. First, we will list and summarize the different schools of thoughts from previous literature describing wettability changes for sandstone and carbonate reservoirs at elevated temperature. Next, we will describe the properties that affect wettability: how they alter wettability and how they are affected by temperature. After that, we will present indications of wettability changes and current wettability measurement techniques.

Following this, case studies describing how wettability change influences reservoir characteristics and field performance during thermal recovery processes will be discussed. The thermal recovery methods included in the case studies were steam flooding, cyclic steam injection, hot water flooding, and in situ combustion. The main and very important take away from this study is that temperature induced wettability change is determined by many possible mechanisms combined together and not by just one or two phenomena occurring simultaneously. Finally, we will propose a reasonable scheme for wettability alteration during dry forward combustion, which needs further investigation.

INTRODUCTION

Understanding formation rock wettability is very important for oil recovery optimization. Wettability is defined as the relative ability of a fluid to spread on a solid surface in the presence of other fluids [1]. The wettability of the fluid/rock system affects the distribution of fluids within a porous medium. The importance of wettability during recovery processes and enhanced oil recovery simulations has been stated previously in several literatures. For instance, a simulation study conducted by Salimi and Bruining [2] concluded that water-flooding results in low recovery in naturally fractured oil wet reservoirs as compared to water wet reservoirs. Certain wettability is favored in various reservoirs and production processes. Roosta et al. [3] stated that neutrally wet condition is favored over strongly oil wet in nonfractured reservoirs. However, not all reservoirs have the original wettability that favors production. The process of wettability alteration is therefore extremely relevant and applicable in these scenarios. Wettability is determined by a combination of many factors such as initial water saturation, saturation history, pH, oil composition and asphaltenes deposition, clay content, and stability of the thin wetting water film on the rock surface [4]. All these properties do not remain constant during production and different recovery processes. In this paper, we will focus on how temperature elevation induced by thermal recovery methods alters some of these properties and therefore changes wettability. Temperature induced wettability alteration had been studied previously, and there were different opinions on how wettability changes in different types of reservoirs.

Temperature Induced Wettability Alteration in Sandstone Reservoirs

Many researchers have carried out experimental work on sandstone cores and studied its wettability with change in temperature. Edmondson [5] reported from his hot water experiment on Berea sandstone that temperature elevation resulted in decreasing relative permeability ratio and residual oil saturation, which indicated that the rock became more water-wet.

Poston et al. [6] conducted a series of dynamic displacement relative permeability measurements on unconsolidated sands at high temperatures. In his study, he found that, for quartz, the residual water saturation and oil-water relative permeability increased and the residual oil saturation decreased with increasing temperature. The observation for unconsolidated sand also concurred with the previously published results as it became more water-wet at high temperature. He also reported that these changes are caused by the change in adhesion tension. Habowski [7] confirmed this observation in his experiment and showed that residual water increased and relative permeability ratio shifted with the increase in temperature. He even specified that the decrease in adhesion tension with increasing temperature is the reason which explains the change in relative permeability ratio.

More work was done by Sinnokrot [8] in which they calculated relative permeability curves from capillary pressure data at elevated temperatures and observed that sandstones became more water-wet as temperature increases. Residual oil saturation and relative permeability to water decreased, whereas irreducible water saturation and the relative permeability to oil increased with increasing temperature. In addition to these consistent results, they also added that hysteresis between drainage and imbibition gradually decreased as temperature is increased. Weinbrandt et al. [9] also carried out experiments with consolidated sands and obtained results similar to Sinnokrot. The shift in relative permeability curves towards higher water saturation with increase in temperature was attributed to the increase in water wetness behavior in the porous medium.

Contrary to the previously published results, some researchers like Wang and Gupta [10] had different observations. They found that the quartz surface became more oil-wet as temperature increases.

Karyampudi [11] also supported the findings of Wang and Gupta. A case study on similar lines was carried out by Blevins et al. [12]. It reviewed field results of steam flooding for heavy oil recovery and found out that the Northern Alberta sandstone reservoir became oil-wet due to steam injection. Recently, Escrochi et al. [13] stated that Berea sandstone became more oil-wet with temperature increase and restored to water-wet with further temperature increase.

These contrasting studies suggest that a comprehensive overview of wettability alteration should be carried out in order to understand its variation with temperature.

Temperature Induced Wettability Alteration in Carbonate Reservoirs

Carbonate reservoirs compared to sandstone reservoirs had very little disagreement between the researchers and most of them published similar observations. Hjelmeland and Larrondo [14] reported that calcite surface changed to water-wet and Lichaa et al. [15] also found similar results. Wang and Gupta [10] as well as Rao [16] published that carbonate became water-wet with temperature increase. Rao [16] confirmed in his study that calcium carbonate precipitation caused by temperature effect was the main reason for a sandstone reservoir not to be altered to oil wet at elevated temperature.

Furthermore, Blevins et al. [12] undertook a case-study and concluded from field data of Qarn Alam field in Oman that carbonate reservoirs became more water-wet with increase in temperature. Also, Al-Hadhrami and Blunt [17] confirmed in their study that carbonate became more water-wet with temperature increment.

Therefore, in general, consensus is observed between the researchers that, with increase in temperature, the carbonate wettability changes to water-wet behavior.

Wettability Is Independent of Temperature

Another school of thought prevails that the increase in irreducible water saturation (S_{wi}) and decrease in residual oil saturation (S_{or}) are independent of temperature and are caused due to the reduction in viscosity ratio.

This hypothesis was supported by the observations of Combarnous and Pavan [18], Hing and Lo Mungan [19], and Sufi et al. [20]. All the aforementioned studies point out that the shift in relative permeability curves is due to reduction in viscosity ratio. As per the case study undertaken by Sufi et al. [20] on the clean unconsolidated Ottawa sand, it was found that the decrease in "Practical" S_{or} was due to the reduction in the viscosity ratio and the "Apparent" increase in S_{wi} was the result of a decrease in the viscous force provided by oil which, in turn, is caused by the viscosity reduction as temperature increases.

TEMPERATURE EFFECTS

As mentioned in the introduction, wettability is determined by a combination of factors such as initial water saturation, saturation history, pH, oil composition and asphaltenes deposition, clay content, and stability of the thin wetting water film on the rock surface [4]. If we want to understand how thermal recovery processes change wettability, we first need to answer the question: how does the elevated temperature caused by thermal processes affect each of the above properties individually?

Rock Composition and Clay Content

Since rocks have different composition, it is expected that they will have different properties, including original wettability. However, temperature elevation plays a role in altering these properties and thus the wettability needs to be investigated.

One of the main differences between sandstone and carbonates is their mineral composition. The main component in sandstone is quartz, which carries negative charge, whereas in carbonates, the main component which is calcium carbonate carries positive charge. In addition, it is determined by Cram [21] that the surface of quartz is weakly acidic and the surface of carbonate is weakly basic. The different charges and pH in sandstone and carbonate result in attraction of different polar components in crude oil and formation brine. For instance, sandstone will tend to attract basic components and the ones carrying positive charge. The different properties of the substance attached to the rock surfaces contribute in defining the wettability. However, experimental results have pointed out that the complex reacting forces between the chemical compounds in reservoir fluid and rock surface might not be simply due to basic attraction mechanisms. Denekas et al. [22] conducted an experiment with crude oil compounds at different acidities and clean cores. The oil fractions were allowed to age on the surface of the rock and results revealed that sandstone was affected by both acidic and basic compounds, whereas carbonates were mainly affected by the basic compounds. The differences in carbonate and sandstone composition and the compounds they attract from formation water lead to different wettability. Treiber et al. [23] concluded that carbonate rocks are more oil-wet than sandstone.

In addition to considering the existing rock properties, we also need to consider the effects of alteration of minerals at elevated temperature, as presented in the paper by Ma and Morrow [24]. During the firing experiment, they observed a permanent change in the structure of quartz, at temperature $573°$ Celsius. There has not been a study conducted on the wettability change due to alteration of quartz crystal orientation; however, Potts and Kuehne [25] conducted firing of Berea sandstone cores and concluded that the core became strongly water-wet after firing. The largest alterations occur within the clay groups. For instance, kaolinite is transformed into smectite at high temperature [26]. Also in the paper by Shaw et al. [27], fine migration was observed during heating, which caused the clay particles to leave the surface of the rock, carrying oil with them and leaving the surface water wet.

Oil Composition

Certain components of crude oil, although present in relatively small amounts can affect the surface properties. For instance, Seifert and Howells [28] stated in their work that carboxylic acid acted like a surfactant at pH greater than 7. The surface active compounds generally contain nitrogen, oxygen, and sulfur. These components are present in the polar fractions of crude oil, which are the heavy components: resin and asphaltenes. This statement is experimentally tested by Denekas et al. [22]. They verified the degree of wettability alteration caused by oil fractions using different molecular weight oil samples. The results indicated that the oil fraction with the heaviest components had the greatest impact on the rock sample and changed it to oil-wet, whereas the other oil fractions had no impact on the rock.

Experiment done by Johansen and Dunning [29] further supported the assumption and demonstrated that asphaltenes played a significant role in altering some of the systems towards oil-wet condition. They compared the resulting wettability of the same crude oil with and without asphaltenes. The results showed that the oil sample without asphaltenes had no impact on water-wet system, whereas the crude oil, which contains asphaltenes, altered the system towards oil-wet system.

Akbarzadeh et al. [30] concluded that Athabasca bitumen asphaltenes precipitation decreased as the temperature was increased. An experimental study of asphaltenes precipitation versus temperature performed by Escrochi et al. [13] theoretically calculated that increase in temperature increases asphaltenes precipitation until the bubble point. With constant pressure, increasing temperature destabilizes the asphaltenes molecules in the crude oil and they form precipitation on the rock until the bubble point, where the difference in solubility parameters of as phaltenes and the mixture has its maximum value. Since as phaltenes are precipitating on the surface of the rock, the rock will behave more oil-wet at this temperature range. Above the bubble point pressure, more as phaltenes are in liquid phase. This means, for saturated and single phase reservoir, increase in temperature will increase as phaltenes solubility so that its precipitation decreases and the reservoir becomes water-wet.

From the above experimental studies, we concluded that the oil composition has significant impact on the system's wettability because crude oil components can act as surface activating agent. Also, the solubility of as phaltenes is a critical factor in determining the systems wettability.

Zeta Potential

As mentioned earlier, Anderson [4] stated that the clay on the surface of the rock absorbed polar compounds in the hydrocarbon such as as phaltenes and resin and made the surface oil-wet. Clay detachment, on the other hand, will reduce the amount of polar molecules on the rock surface, making the surface water-wet. Zeta potential can be used to calculate total potential between two particles in DLVO theory. If zeta potential becomes more negative, it means more repulsive force exists between the particles, whereas if the zeta potential is more positive, there is more attraction between the particles. In the work by Schembre et al. (1998) [31], they developed a function that describes the relationship between zeta potential and temperature for quartz/kaolinite and quartz/illite systems. They concluded, for both systems, that as the temperature increases, zeta potential becomes more negative, which indicates the force is repulsive and leads to clay detachment causing the rock to become more water-wet.

Solid Material Deposition

In the study conducted by Rao [32], he presented an experimental case involving two sides of the same quartz crystal. One side of the crystal was aged with oil and showed strongly oil-wet behavior at all temperatures. However when the other side of the crystal was examined with oil, it exhibited strongly water-wet behavior even after aging in oil and at elevated temperature. The question arose here was whether the properties of the second surface had been altered due to temperature elevation during the test on the first surface. Further inspection of the second surface revealed that the surface was covered with a layer of fine particles, which were identified as salts (mainly calcium carbonate and gypsum) contained in the brine used during the test. Literature review done by Rao [32] indicated that higher temperature results in decreasing solubility of calcium carbonate and causes precipitations. The solid precipitation was the cause of strongly water-wet behavior of the second surface of the quartz crystal. The effect of solid material deposition was later found to be one of the contributing factors that improved oil production rate and decreased water cut in the field [33].

Overall, high temperature will result in precipitation of some minerals that were originally in solution. In the case mentioned in this section, calcium carbonate precipitation caused the sandstone sample to become more water-wet.

Stability of the Thin Wetting Water Film on the Rock Surface

The stability of the thin water film on the surface of the rock is governed by the brine's properties, such as electrolyte concentration and pH. Derjaguin and Churaev [34] observed the effect of different temperature ranges on the stability of thin wetting water films on rock surface. Between 10 and 30°C, the water film on silica decreased dramatically as the temperature increased. Between 30 and 50°C, the decrease in water film thickness with an increase in temperature was not so obvious. And finally at temperature 50°C and above, they observed dramatic water film thickness reduction to almost monolayer of water molecule. They concluded as the temperature increases, the thickness of the wetting water film on silica decreases, which is caused by the breaking of hydrogen bonds. This result indicates that, at higher temperature, the rock surface shifts towards oil-wet tendency due to the thinning of water film.

pH

Generally speaking, the change of pH of a solution at different temperature can be neglected. However, since the wettability of a system partially depends on the pH, we should investigate different mechanisms that describe pH change at elevated temperatures.

Firstly, as mentioned in solid material deposition section, carbonate scales were observed when the brine temperature increases. This precipitation of salt takes out the cations from solution, causing the pH to decrease.

Furthermore, Ramachandran and Somasundaran [35] described the dissolution of silica at high temperature using the following equations:

$$SiO_2 + 2H_2O = H_4SiO_4$$

$$H_4SiO_4 = H^+ + H_3SiO_4^-$$

$$(1)$$

Therefore, we can say that as the temperature increases, the solution became more acidic due to the dissolved silica particles. Schembre and Kovscek (2004) stated in their study that as the pH decreases, zeta potential for kaolinite, illite, and quartz becomes more positive, which contributes to a more oil-wet surface.

Salinity

Salinity is a measurement of the amount of dissolved particles in a solution. In general, we can consider the changes in temperature to have small effects on brine salinity unless precipitation of the salts occurs. Chandrasekhar and Mohanty [36] in their study concluded that lower salinity caused the rocks to become more water-wet, and they listed the following possible mechanisms behind wettability change from other literatures. First, decrease in salinity causes the amount of cations to decrease, which leads to slight increase in pH. Higher pH may cause the generation of in situ calcium soap that will lower the interfacial tension. Secondly, bivalent ions with two bonds are replaced by monovalent ions and this process results in the detachment of the polar hydrocarbon surfactants (resin and asphaltenes), causing the rock surface to become more water-wet. And lastly, like mentioned earlier, decrease in the amount of bivalent and trivalent ions causes the zeta potential to become more negative, which means there is more repulsion between the clay particle and rock surface. The clay particle will be released at the threshold force that breaks the equilibrium and brings the hydrocarbon with them, leaving the surface more water-wet.

Overall, temperature increase will lead to lower brine salinity, which in turn results in more water-wet behavior of the rock surface.

INDICATION OF TEMPERATURE INDUCED WETTABILITY CHANGE

Contact angle measurement is a very simple and accurate indication of the wettability. The most common method used in the laboratory for measuring contact angle is the sessile drop method [4]. The wettability of the surface is determined by measuring the angle formed between the solid surface and liquid at the surface as shown in Figure 1 (adapted from [37]). Figure 1(a) describes a water-wet surface, whereas Figure 1(b) illustrates an oil-wet surface.

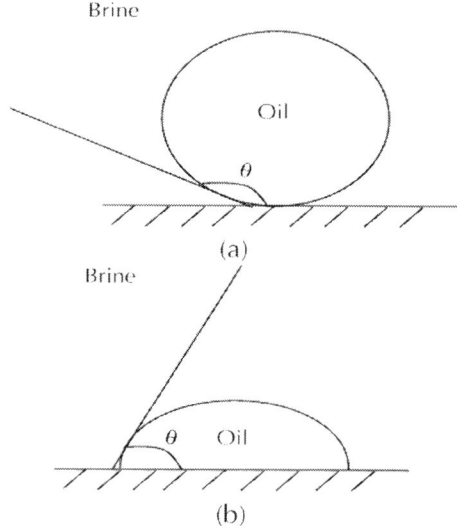

Figure 1: (a) Water-wet surface. (b) Oil-wet surface.

As shown in Figure 1(b), oil tends to adhere to the rock surface in an oil-wet system. Oil molecule will attach and form a film on the surface of the pores and allows the center portion of the small pores to be occupied by water. During recovery processes such as water flooding, water will be channelled through these pores and will leave high residual oil saturation behind. The contact angle of an oil droplet on a surface is known to be affected by the surface tension between rock and the fluids as well as the interfacial tension between different fluid systems in contact with it.

Contact angle generally increases with increase in pressure, but this effect is not that predominant. Variation in temperature alters the contact angle of the fluid as well. This alteration is also dependent on the properties of the solid surface. For example, quartz is observed to be more water-wet compared to calcite at room temperature. Rao et al. [16] carried out contact angle measurements at different temperature conditions using single crystal contact angle and dual drop dual crystal techniques. The results from both the measurement methods indicated that quartz surface became more oil-wet, whereas calcium carbonate surface became water-wet with increase in temperature. This conclusion is contradicting to our existing knowledge of calcium carbonate being an oil-wet system. To understand this discrepancy, it is necessary to distinguish the two main aspects of wettability, namely, spreading and adhesion. In the work by Rao [16], receding contact angle corresponds to the spreading of the droplet of oil on surface, because spreading occurs when oil is displacing existing water causing it to recede. Similarly, the advancing contact angle is measured when water is advancing over a surface which oil was occupying. Normally, contact angle measurement is the measurement of the advancing angle since it closely resembles the nature of fluids flow in the reservoir as oil is produced. However, since these two mechanisms are regarded separately, oil spreading on a surface does not mean that the oil will adhere to the surface and vice versa. In this case, spreading of oil on calcium carbonate is not a direct indication that oil will adhere or form a film instead of water, because oil could spread on top of the water film without the water film being ruptured.

Another physical property indicative of the wettability variation with temperature is relative permeability. It is defined as a measure of the ability of the porous system to conduct one fluid when several fluids are present [38]. Since wettability determines the reservoir fluid's spatial distribution within the rock pores, relative permeability of a system is directly related to wettability. Relative permeability will aid in understanding if the oil phase is going to be easily and efficiently displaced during recovery processes. The viscous forces are temperature dependent and thus the relative permeability curve varies with temperature. For high-tension system, the relative permeability of oil increases, whereas the relative permeability of water decreases with temperature increment at given saturation. These variations result in shifting of the curves to higher water saturation with increase in temperature, indicating a tendency of the rock to become more water-wet. For low-tension system, relative permeability of both water and oil increases with increase in temperature.

Recently conducted experiments by Rajayi and Kantzas [39] indicated that an increase in temperature will lead to lower interfacial tension (IFT) and larger contact angle of the bitumen/water interface. Therefore, the wettability of the system shifts from more water-wet to neutral wettability as temperature increases.

Many researchers observed that the relative permeability values dependent not only on the temperature and interfacial tension effect, but also on certain other physical properties.

Odeh and Cook [40] stated that the decrease in residual oil saturation with increase in temperature is caused by change in viscosity with temperature. Poston et al. (1970) [6] conducted a series of dynamic displacement relative permeability measurements on unconsolidated sands at elevated temperatures. Their results showed for clean quartz an increase in residual water saturation and oil-water relative permeability and a decrease in residual oil saturation with temperature increment. According to them, these changes are caused by the change in adhesion tension. The work of other researchers like Habowski [7], Sinnokrot (1969) [8], and Weinbrandt et al. [9] also concurred with results obtained by Poston et al.

THE EFFECTS OF THERMAL RECOVERY PROCESSES INDUCED TEMPERATURE CHANGE ON WETTABILITY

This following section will show some experiences in the fields which describe how wettability change influences reservoir characteristic and field performance during thermal recovery processes. The cases include hot water and steam injection in pilot project in Qarn Alam field, steam injection in Ghaba North field model, and cyclic steam stimulation (CSS) in pilot project in Elk Cumming Formation. We will also discuss how wettability change affects each zone during dry forward in situ combustion.

Hot Water and Steam Injection in Qarn Alam Field, Oman

Qarn Alam field is a fractured carbonate formation in Shuaiba/Kharaib/Lekhwair reservoir, Oman. The field contains 213 million m³ of heavy oil. The oil density is around 16 API, and the oil viscosity is approximately 220 cp at reservoir condition. Most of the heavy oil is found in the Gharif, Al Khlata, and Haima formation at depth 900–1300 m. The reservoir is oil-wet system with extensive amount of high permeability fractures and low matrix permeability.

The production began in 1975 with the ultimate recovery efficiency under gas lift mechanism only achieving about 2% of stock tank oil initially in place (STOIIP). The other problem is the water cut went up to 95% within two years of production because the reservoir has a strong aquifer support. This problem also increased the water level and left a 40 m of oil rim within the fracture.

Blevins et al. [12] reviewed field studies of steam flooding for heavy oil recovery and found that the Northern Alberta sandstone reservoir became oil-wet due to steam injection and Qarn Alam carbonate reservoir become more water-wet. Based on Al-Adawy and Nandyal [41], Macaulay et al. [42], and Al-Shizawi et al. [43], Qarn Alam field was tested for some thermal recovery projects including hot water and steam injection.

Hot water injection was favored to be the first pilot project in 1985 because this method was expected to enhance the oil recovery by reducing oil viscosity and changing the oil-wet system into water-wet system. During the pilot, hot water was injected at rate 500 m³/d. The bottom-hole temperature increased to 195°C. The residual oil saturation decreased from 88% to 46% across the perforated intervals after hot water flood, which represent significant improvement in oil recovery. However, the test did not observe saturation changes at elevated temperature in the interval above the perforations which made the test results inconclusive and investigation had to be stopped in 1987.

In the study conducted by Tang et al. [44], it was observed that the factor that has the most influence on the oil recovery at elevated temperature in the carbonate reservoir is dominated by imbibition. When temperature increases from 100 to 400°F, the steam injection changes the carbonate reservoir from neutral to strongly water-wet condition, which improves oil recovery by free imbibition. The study by Tang et al. in 2012 [45] also proved that this imbibition potential at high temperature is due to mineral dissolution.

The other proposed method to improve oil recovery in Qarn Alam is Steam Assisted Gas-Oil Gravity Drainage. The pilot project was initiated in 1996. It was forced to a stop in 1997 due to steam-injection plant problem but it was resumed in 1998. The reservoir was heated by steam, and the temperature increased from 50°C to 240°C after 5-month injection.

Laboratory results indicated that, at 50°C, the core was still in oil-wet system which makes the core hard to imbibe with water. After 240°C, the heating induced by steam reduced the oil viscosity to 2 cp. The temperature change also reduced the acid number on carbonate and made the rock water-wet. Therefore, the core imbibes significant amount of water and the gravity drainage rate was improved at elevated temperature. Rao [16] supported this result in his paper stating that carbonate rock becomes water-wet when temperature increases. Steam flood in this field increased total recovery by 11%.

Steam Flooding in Model of Ghaba North Field in Oman

Ghaba North field is carbonate reservoir, with the original wettability being strongly oil-wet. The original oil in place (OOIP) is 119 million m³. The oil density is around 27 API, and oil viscosity is approximately 7 cp. Ghaba North field is a fracture reservoir with 30% average porosity, 10 md matrix permeability, 100 md fracture permeability, and 5–10 m fracture spacing. The fracture spacing of Ghaba North is wider than Qarn Alam. This wide fracture spacing also makes the time scale for heat transport into matrix more significant. Between 1990 and 1995, the field was produced under gas lift, but only 2% recovery was achieved.

Al-Hadhrami and Blunt [17] studied this field and stated that water must overcome a capillary barrier, which was estimated to be 80 KPa to invade the rock matrix and displace the oil in oil-wet rock. Therefore, reducing the capillary barrier by lowering interfacial tension or altering the rock wettability was needed to achieve higher recovery. The change of oil-wet rock to water-wet rock will allow water imbibition into the rock matrix. Therefore, in Ghaba North field, the goal of steam injection is to induce wettability change of the reservoir instead of viscosity reduction.

Roosta et al. discovered in 2009 [3] that steam could reverse the wettability alteration in calcite and quartz surface into water-wet condition by washing out the precipitated asphaltenes molecules from the surface. Das [46] stated the parameters that alter the most in thermal recovery are wettability and capillary pressure.

The clean carbonate rock surface is water-wet initially. When the oil invades the pore, the capillary pressure is exceeded and the water film ruptures. It makes the oil contact the surface directly and the system becomes oil-wet. The temperature elevation by steam injection can change most carbonate reservoir to be more water-wet and increase recovery Morrow et al. (1998) [47]. The steam will heat the rock matrix to induce a wettability change. Then, hot water can spontaneously flow into the matrix and displace oil.

Al-Hadhrami and Blunt [17] developed a 1-dimensional model to analyze the use of thermally induced wettability transition to improve oil recovery in fracture reservoir. From their observation, imbibition will occur over matrix face and create counter current flow. During the steam injection, the imbibition rate is limited by capillary force if the heat is transporting fast. The time for recovery was adjusted with scaling using Ghaba North properties and 30% of recoverable oil is displaced for 20–80 days for counter current imbibition into a matrix block, which indicates the imbibition does not control recovery alone. With advance analysis, the time taken to displace the steam into the matrix with a distance height around 2.5–5 m is estimated between 180 and 740 days. The application in Ghaba North field showed that 30% of oil recovery is displaced in a single matrix block after 700 days of steam injection.

Cyclic Steam Injection

Olsen (1991) [48] stated that wettability shifts towards water-wet behavior as the system temperature increases. He also found that oil saturation decreases during steam flooding. With some of the mechanisms of wettability change stated above, wettability change during cyclic steam stimulation (CSS) has been studied, and the results are debatable. Most scholars believe both sandstone and carbonate will alter to more water wet during CSS process. Hascakir and Kovscek [49] report that the water-oil relative permeability curve shifts to the right at high temperature, which makes the oil production increases drastically in CSS. Schembre et al. (2004) stated wettability change is associated with fine mobilization in the diatomite cores and only occurs at or above a specific temperature. Other cases were also studied by different scholars and results were proven to be similar.

The wettability impact on CSI also has been reported by Karyampudi (1995) [11] through Elk Point heavy oil project. This project was located in Cumming formation of Alberta Field. The field is mainly heavy oil and bitumen. The oil density is 13 API and oil viscosity is 10000 cp at reservoir temperature of 24°C. The primary oil production was 8.5 m³/d with water cut up to 35%. The original oil in place (OOIP) of the section was 7.6 million m³ with the ultimate recovery under primary production 15% OOIP. The main problem of Elk Point was high oil viscosity, poor reservoir quality, and high water saturation.

The CSS pilot was started in 1983. By the end of 1995, there were a total of 27 experimental CSSs. The first four cycles resulted in decrease in oil production and increase in water cuts. The injection and production data analysis showed the wettability of formation around the wellbore reservoir shifted to oil-wet behavior at elevated temperature. The laboratory investigation in 1988 also confirmed this result. Rao and Karyampudi [33] concluded that wettability shifted to oil-wet behavior and reduction in oil relative permeability during CSS was one of the factors responsible for poor performance in one formation. The formation is composed of 87% sand, and at the testing temperature (162−196°C), the sand became oil-wet due to the rupture of wetting film. However, in their paper, they discussed the effect of calcium carbonate deposition due to decreasing solubility at high temperature on wettability and concluded that it prevented the sandstone from changing to more oil wet at high temperature. In addition to the laboratory results, field testing also supported the statement.

Based on Rao and Karyampudi report [33], the wettability control technique using in situ calcium carbonate deposition was implemented in the fifth cycle on 1990. It involved injecting some amount of sodium bicarbonate dissolved in softened water into the steam pipe at the steam generator discharge and some amount of calcium chloride dissolved in hot softened water into the formation in the middle of steam injection.

The result was a slight increase in oil rate and a decrease in S_{or}. However, this technique also triggered an inadvertent fracture event that increases the water cut over to 90%.

From Elk Point thermal pilot performance, Miller and Ramey Jr. (1985) [51] recommended a set of screening guidelines for CSS. The reservoir which is considered for CSS should have oil rates under primary production between 2 and 6 m³/d and water cut under primary production should be lower than 15%.

In Situ Combustion

There was no previous field case that describes the effect of wettability change due to in situ combustion. However, we will have a discussion on the wettability change effect in each zone during in situ combustion process. As shown in Figure 2, there are several zones with different temperatures during in situ combustion process.

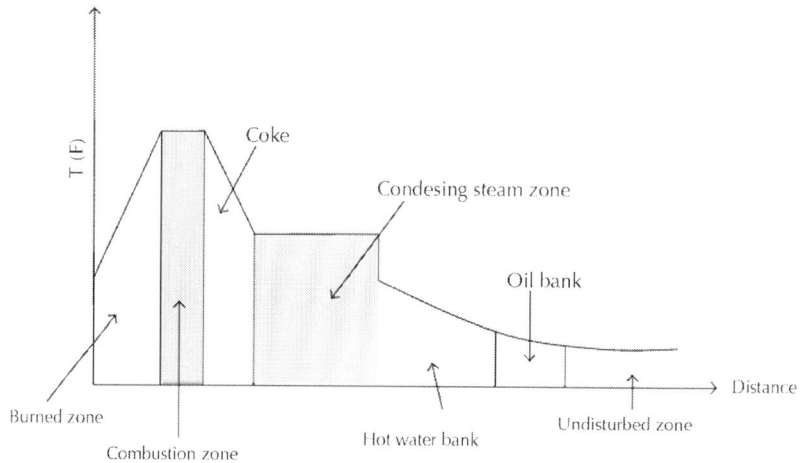

Figure 2: In situ combustion temperature profile for different zones (adapted from [50]).

Technically, temperature can affect relative permeability by changing the wettability or reducing the interfacial tension (IFT) between water and oil. Bardon and Longeron [52] and Haniff and Ali [53] reported that the relative permeability change significantly at IFT below 0.1 dyne/cm². This condition should be observed in the burned zone, combustion zone, and coke zone since all three zones have very high temperature. Therefore, IFT between water and oil can decrease and change the relative permeability for in situ combustion process. On the other hand, the relative permeability will continue to decrease until it approaches zero saturation.

The high temperature also produces water in the gas phase which condenses when reaching a cooler zone and becomes water bank. Butler [54], Castanier and Brigham [55], and Sarathi [50] reported that the steam and water bank that are formed during in situ combustion will influence oil production and act as a hot water injection. In these zones, irreducible water saturation increases and the residual oil saturation decreases smoothly.

The heating effect from coke burning has not reached the undisturbed zone. Therefore, the change of temperature in undisturbed zone is insignificant. The relative permeability curve for this zone will remain similar to the original relative permeability curve.

CONCLUSIONS

- Understanding wettability is very important during oil recovery.
- Wettability is determined by a combination of factors.
- There are discrepancies on how temperature affects rock wettability. The following are possible reasons behind the different thoughts.
    1. Some scholars interpret wettability solely based on relative permeability, but wettability is not the only reason to cause relative permeability to shift.
    2. There is no consistent measurement on wettability. Researchers have been using different methods (Amott test, contact angle measurement, analyzing relative permeability curves, etc.) to evaluate wettability, which causes inconsistency when comparing results.
    3. One has to take into consideration all the factors that affect wettability and not just one or two phenomena occurring separately.
- Clay contents (the amount of each clay type presented in rock) and the degree of clay alteration at elevated temperature need further investigation.
- The wettability alteration mechanism during different EOR techniques, especially in situ combustion, needs more research.

ACKNOWLEDGMENTS

The authors would like to extend their heartfelt gratitude to Dr. Berna Hascakir for her constant motivation and guidance throughout the course of this paper. They also acknowledge the suggestions from all our friends and colleagues towards the betterment of this paper.

REFERENCES

1.   A. G. Mitchell, L. B. Hazell, and K. J. Webb, "Wettability determination: pore surface analysis," in Proceedings of the SPE Annual Technical Conference and Exhibition, New Orleans, La, USA, September 1990

2.   H. Salimi and J. Bruining, "The influence of wettability on oil recovery from naturally fractured oil reservoirs including non-equilibrium effects," in Proceedings of the SPE Latin American and Caribbean Petroleum Engineering Conference, SPE-138366-MS, Lima, Peru, December 2010.

3.   A. B. Roosta, M. F. Escrochi, V. J. Khatibi, V. J. S. Ayatollahi, and M. Schafiee, "Investigating the mechanism of thermally induced wettability alteration," in SPE Middle East Oil and Gas Show and Conference, Bahrain, Bahrain, March 2009.

4.   W. G. Anderson, "Wettability literature survey—part 1: rock/oil/brine interactions and the effects of core handling on wettability," Journal of Petroleum Technology, vol. 38, no. 10, Article ID SPE-13932-PA, pp. 1125–1144, 1986.

5.   T. A. Edmondson, "Effect of temperature on water flooding," Journal of Canadian Petroleum Technology, vol. 4, no. 4, Article ID PETSOC-65-04-09, pp. 236–242, 1965.

6.   S. W. Poston, S. Ysreal, A. K. M. S. Hossain, E. F. Montgomery III, and H. J. Ramey Jr., "The effect of temperature on irreducible water saturation and relative permeability of unconsolidated sands," Society of Petroleum Engineers Journal, vol. 10, no. 2, pp. 171–180, 1970.

7.   E. Habowski, The effect of large temperature changes on relative permeability ratio [M.S. thesis], The Pennsylvania State University, Pennsylvania, Pa, USA, 1966.

8. A. A. Sinnokrot, The effect of temperature on capillary pressure curves of limestone and sandstones [Ph.D. dissertation], Stanford University, 1969.

9. R. M. Weinbrandt, H. J. Ramey Jr., and F. J. Casse, "The effect of temperature on relative and absolute permeability of sandstones," Society of Petroleum Engineers Journal, vol. 15, no. 5, pp. 376–384, 1975.

10. W. Wang and A. Gupta, "Investigation of the effect of temperature and pressure on wettability using modified pendant drop method," in Proceedings of the SPE Annual Technical Conference and Exhibition, SPE 30544, Dallas, Tex, USA, October 1995.

11. R. S. Karyampudi, "Evaluation of cyclic steam performance and mechanisms in a mobile heavy oil reservoir at Elk Point thermal pilot," Journal of Canadian Petroleum Technology, vol. 34, no. 3, pp. 48–56, 1995.

12. T. R. Blevins, J. H. Duerksen, and J. W. Ault, "Light-oil steam flooding—an emerging technology," Journal of Petroleum Technology, vol. 36, no. 8, pp. 1115–1122, 1984.

13. M. Escrochi, M. Nabipour, S. S. Ayatollahi, and N. Mehranbod, "Wettability alteration at elevated temperatures: the consequenses of asphaltene precipitation," in Proceedings of the SPE International Symposium and Exhibition on Formation Damage Control, Lafayette, La, USA, February 2008.

14. O. S. Hjelmeland and L. E. Larrondo, "Experimental investigation of the effects of temperature, pressure, and crude oil composition on interfacial properties," SPE Reservoir Engineering, vol. 1, no. 4, Article ID SPE-12124-PA, pp. 321–328, 1986.

15. P. M. Lichaa, H. Alpuston, J. H. Abdul, W. A. Nofal, and B. F. AlHasan, "Wettability evaluation of a carbonate reservoir rock," in Proceedings of the European Core Analysis Symposium, Paris, France, September 1992.

16. D. N. Rao, "Wettability effects in thermal recovery operations," SPE Reservoir Evaluation & Engineering, vol. 2, no. 5, pp. 420–430, 1999.

17. H. S. Al-Hadhrami and M. J. Blunt, "Thermally induced wettability alteration to improve oil recovery in fractured reservoirs," in Proceedings of the SPE/DOE Improved Oil Recovery Symposium, SPE 71866, Tulsa, Okla, USA, April 2001.

18. M. Combarnous and J. Pavan, "Deplacement par l'eau chaude d'huilesen place dans un milieu poreux," inMe Colloque de l'ARTFP, paper 37, pp. 737–757, Pau, France, September 1968, (French).

19. Y. Hing and N. Lo Mungan, "Effect of temperature on water-oil relative permeabilties in oil-wet and water-wet systems," SPE 4505, 1973.

20. A. H. Sufi, J. H. Ramey Jr., and W. E. Brigham, "Temperature effects on relative permeabilities of oil-water systems," in SPE Annual Technical Conference and Exhibition, New Orleans, Lo, USA, September 1982

21. P. J. Cram, "Wettability studies with non-hydrocarbon constituents of crude oil," Research Report RR-17, Petroleum Recovery Research Inst., 1972.

22. M. O. Denekas, C. C. Mattax, and G. T. Davis, "Effect of crude oil components on rock wettability,"American Institute of Mining, Metallurgical, & Petroleum Engineers, vol. 216, pp. 330–333, 1959.

23. L. E. Treiber, D. L. Archer, and W. W. Owens, "A laboratory evaluation of wettability of fifty oil producing reservoirs," Society of Petroleum Engineers Journal, vol. 12, no. 6, pp. 531–540, 1972.

24. S. Ma and N. R. Morrow, "Effect of firing on petrophysical properties of Berea sandstone," SPE Formation Evaluation, vol. 9, no. 3, pp. 213–218, 1994.

25. D. E. Potts and D. L. Kuehne, "Strategy for alkaline/polymer flood design with berea and reservoir-rock corefloods," SPE Reservoir Engineering, vol. 3, no. 4, pp. 1143–1152, 1988.

26. D. A. Bennion, F. B. Thomas, and D. A. Sheppard, "Formation damage due to mineral alteration and wettability changes during hot water and steam injection in clay-bearing sandstone reservoirs," inProceedings of the SPE Formation Damage Control Symposium, SPE 23783, Lafayette, La, USA, February 1992.

27. J. C. Shaw, P. L. Churcher, and B. F. Hawkins, "Effect of firing on Berea sandstone," SPE Formation Evaluation, vol. 6, no. 1, pp. 72–78, 1991.

28. W. K. Seifert and W. G. Howells, "Interfacially active acids in a California crude oil. Isolation of carboxylic acids and phenols," Analytical Chemistry, vol. 41, no. 4, pp. 554–562, 1969.

29. R. T. Johansen and H. N. Dunning, "Relative wetting tendencies of crude oils by capillarimetric method,"USBM Report RI 5752, 1961.

30. K. Akbarzadeh, O. Sabbagh, J. Beck, W. Y. Svrcek, and H. W. Yarranton, "Asphaltene precipitation from bitumen diluted with n-alkanes," in Proceedings of the Canadian International Petroleum Conference, PETSOC-2004-026-EA, Petroleum Society of Canada, Alberta, Canada, June 2004.

31. M. Schembre, R. Kovscek, and G. Q. Tang, "Effect of temperature on relative permeability for heavy-oil diatomite reservoirs," in Proceedings of the SPE Western Regional Meeting, SPE 93831, Irvine, Calif, USA, March-April 2005.

32. D. N. Rao, "Wettability effects in thermal recovery operations," SPE Reservoir Evaluation & Engineering, vol. 2, no. 5, pp. 420–430, 1996

33. D. N. Rao and R. S. Karyampudi, "Productivity enhancing wettability control technology for cyclic steam process in the Elk Point Cummings formation," Journal of Canadian Petroleum Technology, vol. 38, no. 13, 1999.

34. B. V. Derjaguin and N. V. Churaev, "On the question of determining the concept of disjoining pressure and its role in the equilibrium and flow of thin films," Journal of Colloid and Interface Science, vol. 66, no. 3, pp. 389–398, 1978.

35. R. Ramachandran and P. Somasundaran, "Effect of temperature on the interfacial properties of silicates,"Colloids and Surfaces, vol. 21, pp. 355–369, 1986.

36. S. Chandrasekhar and K. K. Mohanty, "Wettability alteration with brine composition in high temperature carbonate reservoirs," in SPE Annual Technical Conference and Exhibition, vol. 166280 of Proceedings of SPIE, pp. 2416–2432, October 2013.

37. Y. H. Yuan and R. T. Lee, Surface Science Techniques, Springer, Berlin, Germany, 2013.

38. W. G. Anderson, "Wettability literature survey part 5: the effects of wettability on relative permeability,"Journal of Petroleum Technology, vol. 39, no. 11, Article ID SPE-16323-PA, pp. 1453–1468, 1987.

39. M. Rajayi and A. Kantzas, "Effect of temperature and pressure on contact angle and interfacial tension of quartz/water/bitumen system," Journal of Canadian Petroleum Technology, vol. 50, no. 6, Article ID SPE-148631-PA, pp. 61–67, 2009.

40. A. S. Odeh and E. L. Cook, "Discussion—effect of temperature on waterflooding," Journal of Canadian Petroleum Technology, vol. 4, no. 4, p. 242, 1965.

41. M. S. Al-Adawy and M. Nandyal, "Status and scope for EOR development in Oman," in Middle East Oil Show, Society of Petroleum Engineers, November 1991, SPE 21407.

42. R. C. Macaulay, J. M. Krafft, M. Hartemink, and B. Escovedo, "Design of a steam pilot in a fractured carbonates reservoir—Qarn Alam field," in Proceedings of the SPE International Heavy Oil Symposium, vol. 30300 of SPE 30300, Alberta, Canada, June 1995.

43. A. Al-Shizawi, P. G. Denby, and G. Marsden, "Heat-front monitoring in the Qarn Alam thermal GOGD pilot," in Proceedings of the Middle East Oil Show and Conference, SPE 37781, Society of Petroleum Engineers, March 1997.

44. G. Q. Tang, A. Inouye, D. Lowry, and V. Lee, "Recovery mechanism of steam injection in heavy oil carbonate reservoir," in Proceedings of the SPE Western North American Region Meeting, Anchorage, Alaska, USA, May 2011.

45. G.-Q. Tang, A. Inouye, V. Lee, D. Lowry, and W. Wei, "Investigation of recovery mechanism of steam injection in heavy oil carbonate reservoir and mineral dissolution," in Proceedings of the Society of Petroleum Engineers Western Regional Meeting, vol. 153812 of SPE, pp. 457–471, March 2012.

46. S. Das, "Application of thermal recovery processes in heavy oil carbonate reservoirs," in Proceedings of the 15th SPE Middle East Oil and Gas Show and Conference, 2007, SPE 105392.

47. N. R. Morrow, G.-Q. Tang, M. Valat, and X. Xie, "Prospects of improved oil recovery related to wettability and brine composition," Journal of Petroleum Science and Engineering, vol. 20, no. 3-4, pp. 267–276, 1998.

48. D. K. Olsen, "Effect of wettability on light oil steam flooding," Topical Report NIPER-552, USDOE, 1991.

49. B. Hascakir and A. R. Kovscek, "Reservoir simulation of cyclic steam injection including the effects of temperature induced wettability alteration," in Proceedings of the SPE Western Regional Meeting, SPE 132608, Anaheim, Calif, USA, May 2010.

50. P. S. Sarathi, "In-situ combustion handbook—principles and practices," Tech. Rep. DOE/PC/91008-0374, National Petroleum Technology Office, U.S. Department of Energy, Tulsa, Okla, USA, 1999.

51. M. A. Miller and H. J. Ramey Jr., "Effect of temperature on oil/water relative permeabilities of unconsolidated and consolidated sands," Society of Petroleum Engineers Journal, vol. 25, no. 6, pp. 945–953, 1985.

52. C. Bardon and D. G. Longeron, "Influence of very low interfacial tensions on relative permeability," Society of Petroleum Engineers Journal, vol. 20, no. 5, pp. 391–401, 1980.

53. M. S. Haniff and J. K. Ali, "Relative permeability and low tension fluid flow in gas condensate systems," inProceedings of the European Petroleum Conference, SPE-20917-MS, The Hague, Netherlands, October 1990.

54. R. M. Butler, Thermal Recovery of Oil and Bitumen, Prentice Hall, 1991.

55. L. M. Castanier and W. E. Brigham, "Upgrading of crude oil via in situ combustion," Journal of Petroleum Science and Engineering, vol. 39, no. 1-2, pp. 125–136, 2003.

Laboratory Study on the Potential EOR Use of HPAM/VES Hybrid in High-Temperature and High-Salinity Oil Reservoirs

Dingwei Zhu[1,2], Jichao Zhang[3], Yugui Han[3], Hongyan Wang[3], and Yujun Feng[1,3,4]

[1]Chengdu Institute of Organic Chemistry, Chinese Academy of Sciences, Chengdu 610041, China

[2]University of Chinese Academy of Sciences, Beijing 100039, China

[3]EOR Laboratory, Geological Scientific Research Institute, Shengli Oilfield Company of SINOPEC, Dongying 257013, China

[4]State Key Laboratory of Polymer Materials Engineering, Polymer Research Institute, Sichuan University, Chengdu 610065, China

ABSTRACT

Polymer flooding represents one of the most efficient processes to enhance oil recovery, and partially hydrolyzed polyacrylamide

(HPAM) is a widely used oil-displacement agent, but its poor thermal stability, salt tolerance, and mechanical degradation impeded its use in high-temperature and high-salinity oil reservoirs. In this work, a novel viscoelastic surfactant, erucyl dimethyl amidobetaine (EDAB), with improved thermal stability and salinity tolerance, was complexed with HPAM to overcome the deficiencies of HPAM. The HPAM/EDAB hybrid samples were studied in comparison with HPAM and EDAB in synthetic brine regarding their rheological behaviors and core flooding experiments under simulated high-temperature and high-salinity oil reservoir conditions (T: 85°C; total dissolved solids: 32,868 mg/L; $[Ca^{2+}] + [Mg^{2+}]$: 873 mg/L). It was found that the HPAM/EDAB hybrids exhibited much better heat- and salinity-tolerance and long-term thermal stability than HPAM. Core flooding tests showed that the oil recovery factors of HPAM/EDAB hybrids are between those of HPAM and EDAB. These results are attributed to the synergistic effect between HPAM and EDAB in the hybrid.

INTRODUCTION

Among all the chemically enhanced oil recovery (CEOR) processes, polymer flooding represents one of the most efficient methods to produce residual oil from depleted and water-flooded reservoirs [1, 2]. In this process, the increased viscosity of the displacing fluid by the added water-soluble polymer will improve the mobility ratio between the injected fluid and the reservoir oil, mobilizing the capillary trapped water-flooded oil in the secondary stage, leading to better vertical and areal sweep efficiencies and thus higher oil recovery efficiencies [2]. In China, around 13 million tons of oil is produced additionally per year by this chemical flooding technique.

Partially hydrolyzed polyacrylamide (HPAM) is the most widely used oil displacement agent and has been successfully employed in polymer flooding worldwide [2–4]. As an anionic polyelectrolyte, HPAM is easily to be dissolved in water and shows strong thickening ability in fresh water at relatively lower dosage. Nevertheless, the notorious congenital drawbacks of HPAM also limit its applications in hostile environment, in particular, the high-temperature, high-salinity, and low-permeability oil reservoirs [5], such as the Class III reserve of Shengli Oil Field in China where the temperature is above 85°C and the salinity (total dissolved solids, TDS) is higher than 30,000 mg/L in which the total amount of Ca^{2+} and Mg^{2+} exceeds 800 mg/L. In such a harsh environment, the interaction of metal ions in the oil field brines largely shields the mutual repulsion from the carboxylic groups along the HPAM skeleton, leading the polymer coils to collapse, decreasing hydrodynamic volume, and thus ultimately lowering solution viscosity [2]. Efficiency loss of HPAM aqueous solution at elevated temperature becomes further serious as more amide groups undergo extensive hydrolysis into carboxylic characters, and the resulting hydrolyzed products precipitated when contacting Ca^{2+} and Mg^{2+} [6], commonly present in oil reservoir brines or hard water.

Another major limitation of HPAM is its flow-induced mechanical degradation. As a synthetic polymer, HPAM is intrinsically linear, flexible, high-molecular weight (generally higher than ten million), and highly polydispersible in molecular weight (polydispersity index normally between 2 and 3). When it is mixed and dissolved in tanks, or it is passing through chokes, pipes, valves, nozzles, pumps, perforations near wellbore, or the pore throat in the porous media, HPAM chains are subjected to both shear flows and elongation flows [7]. In pure shear flows where the shear rate is perpendicular to the flow direction, HPAM molecules rotate and are thus subjected to periodic extensional and compressional stresses. As shear rate increases, hydrodynamic stresses increase and their effects are nonnegligible compared with those of Brownian motion when rotation time becomes smaller than the longest rotational relaxation time. In pure elongational flows where the elongation rate is parallel to flow direction, the effective deformation that occurs in the flow direction may be very large for the flexible HPAM and a full stretching is achieved only if the macromolecule remains in the elongational flow over a sufficiently long time. When submitted to shear, macromolecules do not stretch significantly, whereas when submitted to extension, they can elongate drastically and break. The first consequence of macromolecule stretching in elongational flows is a strong increase in viscosity. For high-molecular-weight HPAM used in EOR, the elongational viscosity may be as high as 10^4 times the zero-shear-rate viscosity. The second consequence is that the macromolecule is subjected to an internal tensile. It has been shown that elongation force could possibly be high enough to exceed the carbon-carbon bond force and thus may cause chain breakage, that is, mechanical degradation of HPAM chains [7]. This degradation process essentially breaks the larger molecules up into smaller fragments and thus changes the molecular weight distribution of the HPAM, which hinders the efficiency of the EOR technique [8, 9].

Facing these severe challenges, two options were naturally used to improve the properties of HPAM in high-temperature and high-salinity environment; one is introducing thermostable and more salt-tolerant monomer or groups such as SO_3^- onto the HPAM backbone [10], and the second is increasing molecular weight of HPAM (maximum 35×10^6 g/mol to date) to get higher viscosity retention. However, the main portion in the first case is still the acrylamide segment which shows poor long-term thermal stability and salt tolerance, and in the second case, the increased molecular weight of HPAM will bring about easier mechanical degradation [7] and the plugging of the smaller pore throat in the low-permeability oil reservoirs. Thus, we shifted our attention to seek other alternatives with better salt tolerance and thermal stability, as well as improved mechnical stability, and we found recently that viscoelastic surfactant (VES) is one of such choices amongst others.

Viscoelastic surfactants are formed from the entanglement of worm-like micelles (WLMs) by certain surfactants in the presence of an organic or inorganic hydrotrope [11]. WLMs are long flexible aggregates of surfactant molecules in aqueous solution, and above a surfactant concentration threshold, WLMs entangle into a transient network that is constantly breaking and reforming, for which they are also referred to as "living" or "equilibrium" polymers [12, 13]. The entanglement of WLMs imparts strong thickening ability of water and remarkable viscoelastic properties to the solution. These peculiar rheological properties of VES, including reversible breaking and formation, viscoelastic response analogous to polymer solutions, and strong viscosifying ability, may furnish them to be applied in CEOR process.

Up to date, most of the VES systems are composed of cationic surfactants whose hydrophobic tail is normally shorter than C18. The strong adsorption between these cationic amphiphiles and negatively charged sandstone, as well as high dosage needed to thicken injection fluid, impeded their practical use in EOR process. To overcome these deficiencies of cationic VES, we recently developed ultra-long chain particularly C22-tailed zwitterionic [14–19], anionic [20, 21], nonionic [22] surfactants which are capable of forming WLMs at much lower concentration; that is, the overlapping concentration (C*) is much lower than those of C16 cationic counterparts. Apart from these advantages, the C22-tailed zwitterionic surfactants—3-(nerucamidopropyl-N,N-dimethyl ammonium) propane sulfonate (EDAS) [14–17, 19] also show additional advantages such as insensitivity to inorganic salt, sufficient stability over the whole pH range [14], and long-term thermal stability [15]. These unique properties of EDAS make it suitable for potential use in high-temperature and high-salinity reservoirs. Nevertheless, the betaine agent to prepare EDAS, 1,3-propanesultone, is extremely poisonous and carcinogenic, restricting the scale-up production and industrial use of EDAS in tertiary oil recovery.

Thus, in this work, a newly synthesized carboxylate counterpart of EDAS, erucyl dimethyl amidopropyl betaine (EDAB), was chosen for the study of its possible use in chemical EOR process. Compared with its saturated-chain counterparts, EDAB shows improved water solubility and much lower critical micelle concentration [18]. However, we found that the oil recovery factor from single VES solution is quite low. Therefore, HPAM was used to complex with EDAB to form hybrids which were systematically examined to see their potential in CEOR process. Reported here are the rheological behaviors and preliminary core flooding results of these hybrids under simulated high-temperature and high-salinity oil reservoirs of Shengli oil Field. For comparison, HPAM and EDAB were also tested under the same experimental conditions.

EXPERIMENTAL

Materials

HPAM (Scheme 1) with viscosity-averaged molecular weight of 1.2×10^7 g/mol was obtained from Beijing Hengju Chemical Group Co., Ltd. The hydrolysis degree and active content of this polymer are 24% and 89.9%, respectively. EDAB (Scheme 1) was prepared by the reaction of the corresponding fatty acids with N,N-dimethyl-1,3-propanediamine, followed by quaternization with sodium chloroacetate of the obtained intermediates [18]. All other chemicals used are analytical grade, and the water used is doubly distilled.

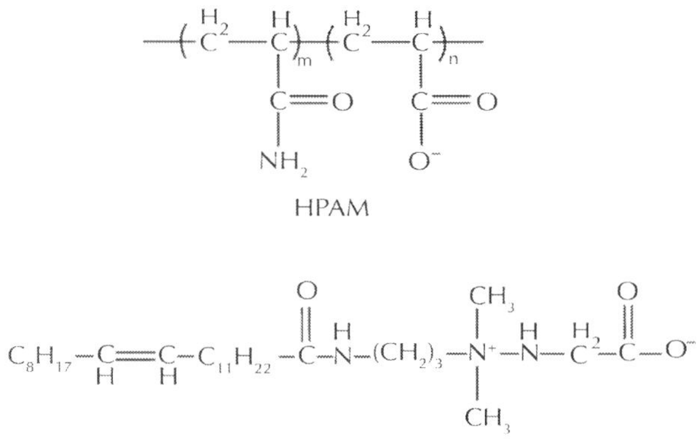

HPAM

EDAB

Scheme 1: Molecular structures of HPAM and EDAB used in this work.

Preparation of Hybrid Samples

The HPAM or EDAB stock solutions with concentration of 0.3% were separately prepared by dissolving designed amount of powders into synthetic brine (TDS: 32,868 mg/L; $[Ca^{2+}] + [Mg^{2+}]$: 873 mg/L) with gentle stirring at room temperature. The TDS and hardness of the brine are close to those of connate water in the Reserve III of Shengli Oil Field. The hybrid samples were prepared by mixing EDAB solution with HPAM solution following the designed recipe and gently stirred for 3 days, and then left to stand for 1 day prior to test. The sample code "PxEy" refers to the concentration of the two chemicals in the mixture (Table 1).

Table 1: The nomenclature and compositions of chemicals used in this work

Sample name	HPAM (wt%)	EDAB (wt%)
P30	0.3	0
E30	0	0.3
P25E5	0.25	0.05
P20E10	0.20	0.10
P15E15	0.15	0.15
P5E25	0.05	0.25

Rheology Test

All rheological measurements were carried out on a Physica MCR301 (Anto-Paar, Austria) rotational rheometer equipped with concentric cylinder geometry CC27. The radii of the measuring bob and the measuring cup are 13.33 and 14.46 mm, respectively. The temperature was controlled by a Peltier system that provides fast and precise adjustment of the temperature during heating. A solvent trap was used to prevent evaporation of the solvents during measurement.

The apparent viscosity (η_{app}) of the samples was measured at 85°C with a fixed shear rate (γ) of 7.34 s^{-1}. Steady shear viscosity was recorded during temperature scans going from 20 to 85°C (heating rate = 2°C/min). The elastic and storage moduli of the samples were registered during the angular frequency scans from 0.05 to 100 rad·s^{-1} at 85°C with the same rheometer.

Long-Term Thermal Stability Test

0.05% $Na_2S_2O_3$ is added into all the samples as an oxygen scavenger, and the sample solutions were distributed into 120-mL glass bottles, and then sealed with a cover, followed by placing them into an oven and aged at 85°C. At consecutive time intervals, the samples were taken out for viscosity monitoring.

Core Flooding Test

The core flooding tests were performed under the simulated high-temperature and high-salinity oil reservoirs in Shengli Oil Field of China and followed a previously reported procedure [23]. A steel cylinder with 2.5 cm in inner diameter and 25 cm in height was packed with several sizes of silica sand in model. The porosity of sandstone core was 32.6%. The sand pack was initially saturated with the synthetic brine, followed by injecting the dehydrated Shengli oil to ~70%. The density of dehydrated Shengli oil was 0.95 g/cm^3 and the viscosity was 40.1 mPa·s. The core was then injected with the synthetic brine until the water content was higher than 98% in the output fluid, and the injection of 30% pore volume (PV) of the sample solution was followed. The total oil recovery and the oil recovery by water flooding were calculated, respectively, and the difference between them was the tertiary oil recovery factor by polymer or polymer/surfactant hybrid flooding. The injection rate of the sample was maintained at 0.23 mL/min, and the cylindrical sand pack was placed in a chamber and heated at 85°C throughout the test.

RESULTS AND DISCUSSION

Rheological Behaviors of HPAM/EDAB Hybrid Samples

It is well accepted that apparent viscosity of polymer solution plays a crucial role in displacing less viscous oil in the EOR process [24]. Therefore, it is necessary to investigate the viscous properties of the hybrid samples before flooding test.

Exhibited in Figure 1 are the apparent viscosities of different samples plotted against temperature. As expected, η_{app} of HPAM solution (P30) decreases continuously upon increasing temperature, with a general trend following Arrhenius laws. The final η_{app} at 85°C is just one-third of the original value at room temperature, which implies that the HPAM solution shows pure thermothinning behavior. However, η_{app} of EDAB solution (E30) decreases sharply before 25°C, following an unstable increase until 56°C, after which the viscosity drops again. Typically, its viscosity between 70 and 85°C remains unchanged. But if one compares the η_{app} of EDAB at 30 and 85°C, you will find that there is no change between them. This means that EDAB is less thermosensitive than HPAM. Interestingly, the HPAM/EDAB hybrid samples such as P15E15 and P10E20 exhibit more similar behavior to EDAB rather than HPAM solution. Both of the two hybrid samples show inverted "V" shape in viscosity-temperature curves, all of which have a viscosity peak 55°C, analogous to that of EDAB. But it is worth noting that the higher the polymer fraction, the higher the viscosity peak. It is also noteworthy that the hybrid samples show stronger thermo-thinning behaviors after the viscosity maximum compared to EDAB.

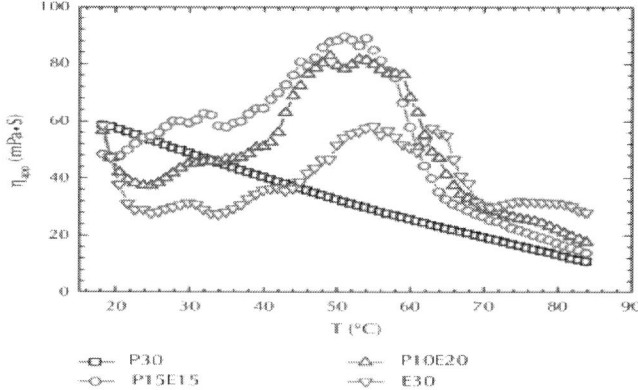

Figure 1: Effect of temperature on apparent viscosity of the samples (TDS = 32,868 mg/L, [Ca^{2+}] + [Mg^{2+}] = 873 mg/L, = 7.34 s^{-1}, heating rate: 2°C/min).

To pick out the optimized formulation used for oil displacement under the simulated oil reservoir condition, it is utmost important to compare all the hybrid samples, as well as the pure HPAM and EDAB solutions, at the target temperature. Summarized in Table 2 are the apparent viscosities of all the samples. It is quite surprising that there is no direct correlation between viscosity and the content of EDAB or HPAM. Although the viscosity of 0.3% EDAB is 10 mPa·s higher than that of 0.3% HPAM, the apparent viscosity of the 0.3% hybrid samples decreases first followed by continuous increase upon increasing the content of EDAB in the hybrid samples. Exceptionally, the sample P10E20 composed of 0.10% HPAM and 0.20% EDAB displays the apparent viscosity as high as 30.61 mPa·s, while P5E25 with 0.05% HPAM and 0.25% EDAB possesses viscosity of only 18.29 mPa·s, and P15E15 only has 11.25 mPa·s. In the following studies, the two hybrid samples, P15E15 and P10E20, will be employed for long-term thermostability and core flooding studies.

Table 2: Apparent viscosity of the samples measured at simulated high-temperature and high-salinity oil reservoirs (TDS = 32,868 mg/L, [Ca^{2+}] + [Mg^{2+}] = 873 mg/L, T = 85°C, = 7.34 s^{-1})

Sample	P30	P25E5	P20E10	P15E15	P10E20	P5E25	E30
ηapp (mPa·s)	17.93	11.34	8.59	11.25	30.61	18.29	28.04

In order to distinguish whether HPAM undergo microstructural changes with the addition of EDAB, dynamic rheological experiments were performed at 85°C as well. Figure 2 shows the plots of storage modulus (G') and loss modulus (G") as a function of oscillatory shear frequency (W) for the HPAM/EDAB hybrid samples. For the HPAM sample (P30), both moduli are strong functions of shear frequency over the entire frequency. When EDAB is added, the G' and G" of the hybrid samples P15E15 and P10E20 remain nearly unchanged, especially at high frequencies. But it is worth noting that there are crossovers between G' and G" for all the samples, indicative of the formation of network structures in the solution. These results suggest that HPAM does not undergo microstructural changes when EDAB is added and it still has the ability of increasing the oil recovery factor. Therefore, the hybrid samples have great potential to enhance oil recovery from high-temperature and high-salinity oil reservoirs because EDAB shows excellent heat and salinity tolerance.

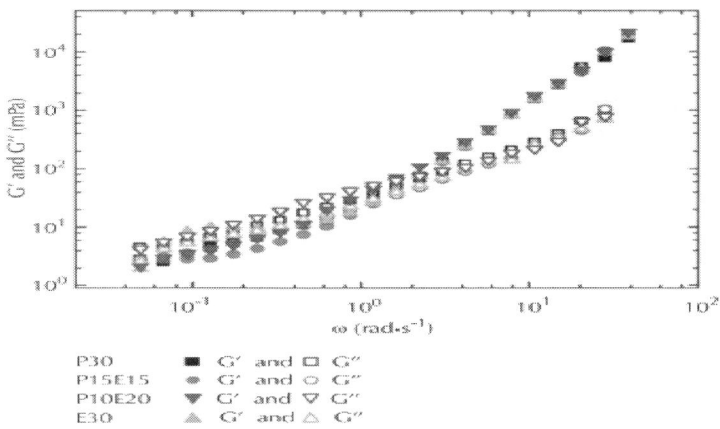

Figure 2: Storage modulus (G') and loss modulus (G") plotted as a function of angular frequency (W) for the samples (TDS = 32,868 mg/L, $[Ca^{2+}] + [Mg^{2+}]$ = 873 mg/L, T = 85°C).

Long-Term Thermal Stability

The remaining viscosity at high temperature represents a primary criterion for any chemicals to be used in hostile environment. For instance, the continuous aging of the displacing fluid is detrimental to tertiary oil recovery [25], particularly in high-temperature oil reservoirs. Therefore, high-temperature aging is crucial to EOR chemicals, and long-term thermal stability experiment of HPAM/EDAB hybrid samples is inescapably necessary.

Figure 3 shows the variation of η_{app} as a function of aging time for the HPAM/EDAB hybrid samples and sole HPAM or EDAB after aging at 85°C. One can clearly find a sharp reduction in η_{app} for HPAM (P30) after aging: the initial η_{app} (0 d) is 17.93 mPa·s, but is only 4.88 mPa·s left for the same sample after 3 days of aging! There is also a sharp reduction in η_{app} for P10E20 in 6 days: its η_{app} decreases to 18.11 mPa·s, and 41% of loss in viscosity. However, after more than 25 days of aging, the η_{app} of P10E20 begins to increase other than decrease. Unlike P30 and P10E20, the η_{app} of P15E15 maintains a constant value regardless of aging time, showing improved long-term thermal stability over HPAM. Meanwhile, as shown in Figure 3, EDAB (E30) exhibits a much better thermal stability than HPAM after 66 days of aging. These results clearly demonstrate that the ultra-long-chain zwitterionic surfactant EDAB possesses much better long-term thermal stability over HPAM, and the addition of EDAB into HPAM solutions will improve the thermal stability of HPAM solutions.

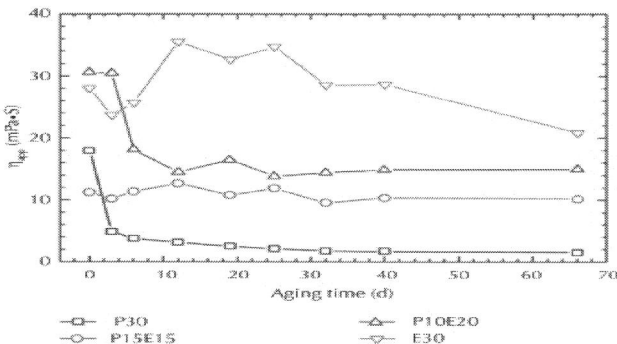

Figure 3: Long-term thermal stability of HPAM/EDAB hybrid samples in comparison with EDAB and HPAM (TDS = 32,868 mg/L, [Ca^{2+}] + [Mg^{2+}] = 873 mg/L, = 7.34 s^{-1}). Both the aging and measuring temperature is 85°C.

Oil Displacement Test

To our knowledge, there are no published data concerning chemical flooding using HPAM or EDAB under the simulated high temperature and salinity oil reservoir conditions. Although the HPAM/EDAB hybrid samples show promising potential for chemical EOR in high-temperature and high-salinity oil reservoirs, no core flooding test so far has been performed yet.

Plotted in Figure 4 are the recovery factors, water cut, and flooding pressure as a function of injected volume of the sample solutions under the simulated high-temperature and high-salinity oil reservoir environment. Table 3 shows core parameters, displacement process, and the results of these oil displacement tests. It was found that 15.85% oil recovery factor by HPAM (P30) flooding was obtained at these conditions (Figure 4(a)), whereas the oil recovery factor of EDAB (E30) was only 1.90% (Figure 4(b)). As shown in Figures 4(c) and 4(d), the oil recovery factor of P15E15 was 10.20% and the factor of P10E20 was 7.10%; both of them are smaller than that of HPAM, but much higher than that of EADB. The main reason to get a higher oil recovery factor with the hybrid samples can be ascribed to strengthened micelles by the added HPAM long chains which are not as easy as that of micelles to be disrupted. It is worth emphasizing that the synergistic effect between HPAM and EDAB enables the hybrid sample to show an excellent long-term thermal stability and get a relatively high oil recovery factor. Under long-term propagation in the simulated high-temperature and high-salinity oil reservoirs, the apparent viscosity of the HPAM/EDAB hybrid sample is relatively stable, thus effectively improving water-to-oil mobility ratio to make more oil produced.

Table 3: Core parameters, displacement process, and recovery factors

Core no.	Permeability (mDarcy)	Pore volume (cm3)	Saturated oil (cm3)	Slug	Slug injected (PV)	Water flooding recovery (%)	Ultimate recovery (%)	Enhanced oil recovery (%)
1	1059	53.4	43.0	P30	0.3	55.10	70.95	15.85
2	1480	50.6	42.0	P15 E15	0.3	60.85	71.05	10.20
3	1496	50.0	41.0	P10 E20	0.3	45.10	52.20	7.10
4	1494	49.0	41.0	E30	0.3	35.95	37.85	1.90

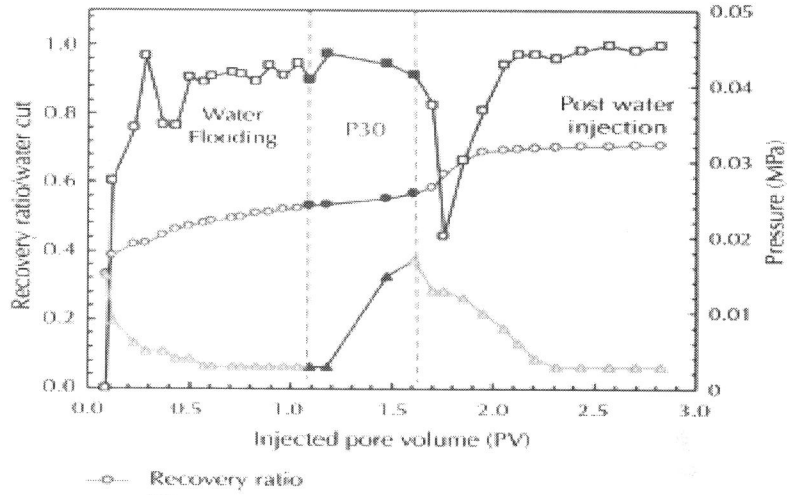

(a)

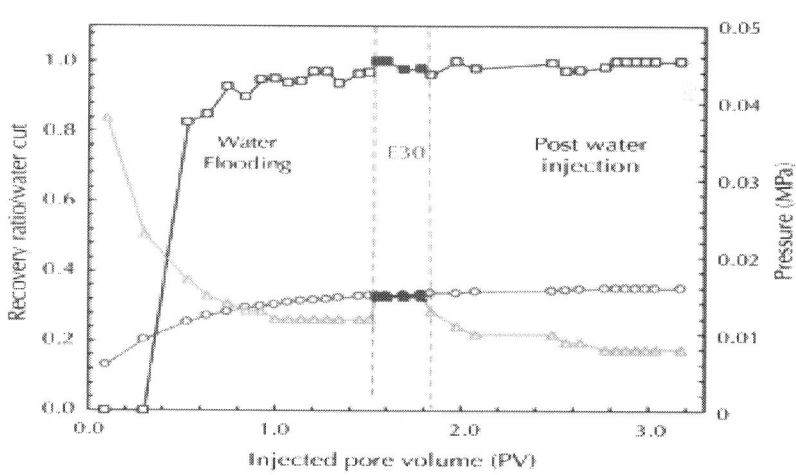

(b)

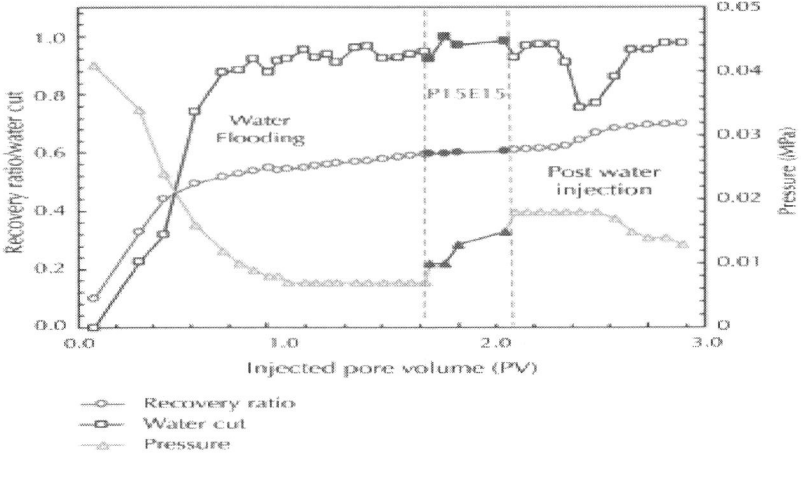

(c)

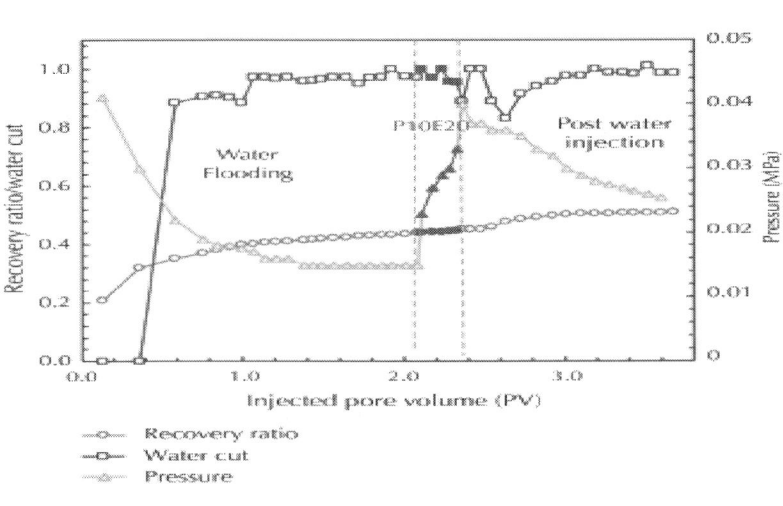

(d)

Figure 4: Recovery factor, water cut, and flooding pressure plotted as a function of injected volume of the samples: (a) P30, (b) E30, (c) P15E15, and (d) P10E20 (TDS = 32,868 mg/L, [Ca^{2+}] + [Mg^{2+}] = 873 mg/L; injected volume = 30% PV; injected rate = 0.23 mL/min).

In spite of its poor long-term thermal stability, the 0.3% HPAM solution can still enhance oil recovery to a higher extent in the relative short period (10 hours) of oil displacement test. On the contrary, the oil recovery factor of EDAB is extremely low though it has much stronger thickening ability and much better long-term thermal stability under the simulated high-temperature and high-salinity oil reservoirs. The poor oil recovery efficiency may be attributed to the collapse of the micelles upon contacting oil, and the disassembled micelles by the shear and elongation force at the pore throats cannot be recovered when they reach the next throat. However, such a hypothesis should be further verified quantitatively in future studies.

The propagation of the different chemicals in porous media is schematically illustrated in Figure 5. In the flooding process, all the samples are subjected to shear and stretch in the pores of porous reservoir media. In this case, WLMs are destroyed and cannot form micelles in a short time (Figure 5(a)). However, HPAM molecular chains orient the flow direction and reform polymer coils after going through the porous media (Figure 5(b)). For the HPAM/EDAB hybrid samples, HPAM assists WLMs in going across the pores and prevents them from being destroyed (Figure 5(c)). Therefore, the oil recovery factor of the HPAM/EDAB hybrid sample is smaller than that of HPAM, but much higher than that of EADB.

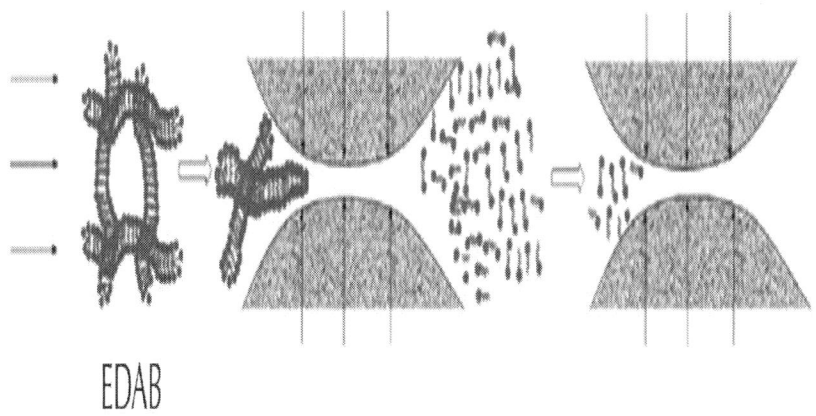

EDAB

(a)

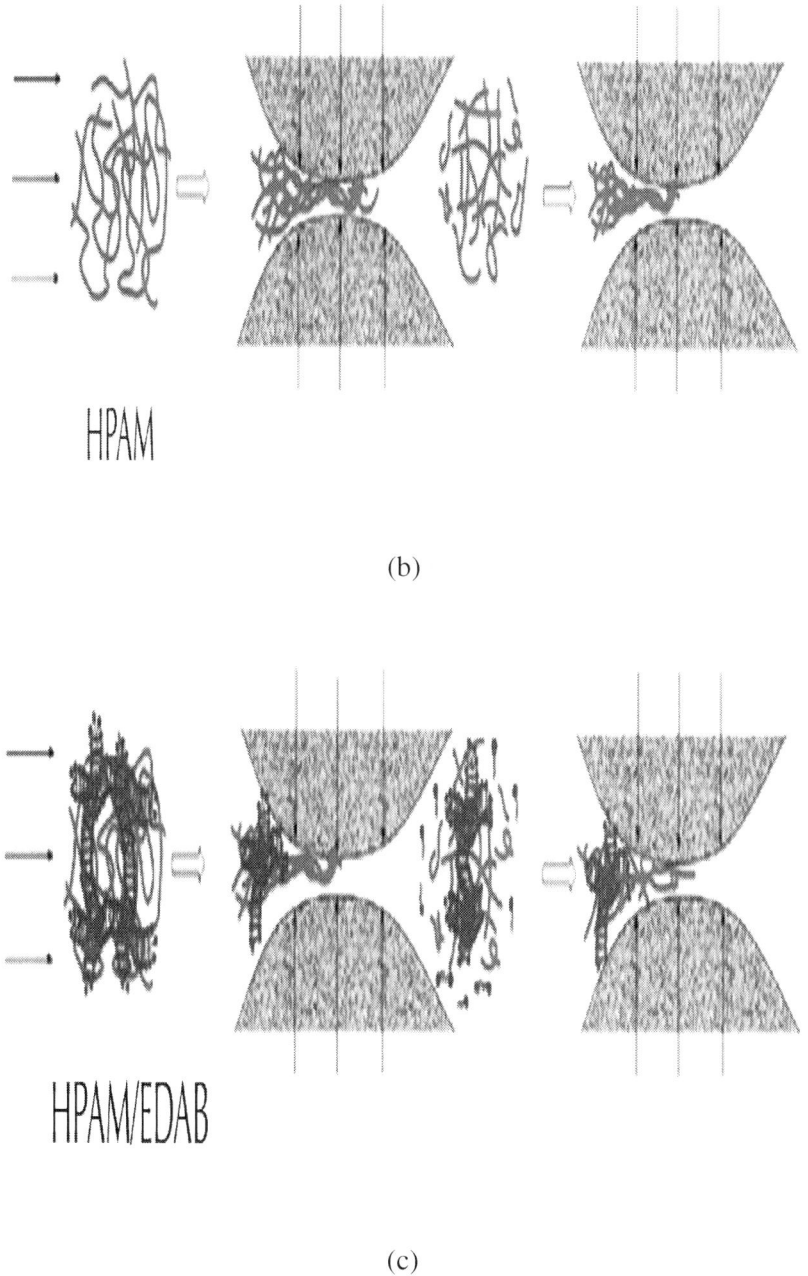

HPAM

(b)

HPAM/EDAB

(c)

Figure 5: Schematic description of the chemical flooding processes for the samples: (a) EDAB, (b) HPAM, and (c) HPAM/EDAB hybrid.

CONCLUSION REMARKS

The rheological behaviors of partially hydrolyzed polyacrylamide/amidobetaine surfactant hybrid and core flooding experiments under simulated high temperature and salinity oil reservoirs conditions were preliminarily examined. It was found that the HPAM/EDAB hybrid samples (P15E15 and P10E20) exhibited improved salt tolerance and long-term thermal stability. In addition, 10.20% and 7.10% of oil recovery factors were achieved from the HPAM/EDAB hybrid samples in the laboratory core flooding test. Along with their improved salt tolerance and long-term thermal stability, these hybrid samples show great potential to enhance oil recovery from hostile oil reservoir environment. However, it is necessary to further investigate the mechanism of the HPAM/EDAB hybrid and the optimum ratio between HPAM and EDAB for economic consideration.

ACKNOWLEDGMENTS

The authors are grateful to the financial support from Shandong Provincial government through the "Taishan Scholar" project, Science and Technology Department of Sichuan Province (2012NZ0006, 2010JQ0029), as well as Natural Science Foundation of China (21173207, 21273223), and Chinese Academy of Sciences.

REFERENCES

1. S. Thomas, "Enhanced oil recovery—an overview," Oil & Gas Science and Technology—Revenue d'IFP Energies Nouvelles, vol. 63, no. 1, pp. 9–19, 2008.

2. K. S. Sorbie, Polymer-Improved Oil Recovery, Baca Raton, Fla, USA, 1991.

3. J. C. Jung, K. Zhang, B. H. Chon, and H. J. Choi, "Rheology and polymer flooding characteristics of partially hydrolyzed polyacrylamide for enhanced heavy oil recovery," Journal of Applied Polymer Science, vol. 127, no. 6, pp. 4833–4839, 2013.

4. Q. Chen, Y. Wang, Z. Lu, and Y. Feng, "Thermoviscosifying polymer used for enhanced oil recovery: rheological behaviors

and core flooding test," Polymer Bulletin, vol. 70, no. 2, pp. 391–401, 2013.

5.   Y. Wang, Z. Lu, Y. Han, Y. Feng, and C. Tang, "A novel thermoviscosifying water-soluble polymer for enhancing oil recovery from high-temperature and high-salinity oil reservoirs," Advanced Materials Research, vol. 306-307, pp. 654–657, 2011.

6.   A. Zaitoun and B. Poitie, "Limiting conditions for the use of hydrolysed polyacrylamides in brines containing divalent ions," in Proceedings of the SPE International Oilfield and Geothermal Chemistry Symposium, Paper SPE 11785, Denver, Colo, USA, June 1983.

7.   G. Chauveteau and K. S. Sorbie, "Mobility control by polymers," in Basic Concepts in Enhanced Oil Recovery Process, M. Bavière, Ed., vol. 33 of Critical Reports on Applied Chemistry, pp. 44–87, Elsevier, London, UK, 1991.

8.   A. Dupas, I. Hénaut, J. F. Argillier, and T. Aubry, "Mechanical degradation onset of polyethylene oxide used as a hydrosoluble model polymer for enhanced oil recovery," Oil & Gas Science and Technology—Revenue d'IFP Energies Nouvelles, vol. 67, no. 6, pp. 931–940, 2012.

9.   C. Noïk, P. Delaplace, and G. Müller, "Physico-chemical characterics of polyacrylamide solutions after mechanical degradation through a porous medium," in Proceedings of the SPE International Symposium on Oilfield Chemistry, Paper SPE 28954, San Antonio, Tex, USA, February 1995.

10. G. Dupuis, D. Rousseau, R. Tabary, J.-F. Argillier, and B. Grassl, "Hydrophobically modified sulfonated polyacrylamides for IOR: correlations between associative behavior and injectivity in the diluted regime," Oil & Gas Science and Technology—Revenue d'IFP Energies Nouvelles, vol. 67, no. 6, pp. 903–920, 2012.

11. J. Yang, "Viscoelastic wormlike micelles and their applications," Current Opinion in Colloid & Interface Science, vol. 7, no. 5-6, pp. 276–281, 2002.

12. C. A. Dreiss, "Wormlike micelles: where do we stand? Recent developments, linear rheology and scattering techniques," Soft Matter, vol. 3, no. 8, pp. 956–970, 2007.

13. Z. Chu, C. A. Dreiss, and Y. Feng, "Smart wormlike micelles," Chemical Society Review, vol. 42, pp. 7174–7203, 2013.

14. Z. Chu, Y. Feng, X. Su, and Y. Han, "Wormlike micelles and solution properties of a C22-tailed amidosulfobetaine surfactant," Langmuir, vol. 26, no. 11, pp. 7783–7791, 2010.

15. Z. Chu, Y. Feng, H. Sun et al., "Aging mechanism of unsaturated long-chain amidosulfobetaine worm fluids at high temperature," Soft Matter, vol. 7, no. 9, pp. 4485–4489, 2011.

16. Z. Chu and Y. Feng, "Amidosulfobetaine surfactant gels with shear banding transitions," Soft Matter, vol. 6, no. 24, pp. 6065–6067, 2010.

17. Z. Chu and Y. Feng, "A facile route towards the preparation of ultra-long-chain amidosulfobetaine surfactants," Synlett, no. 16, pp. 2655–2658, 2009.

18. D. Feng, Y. Zhang, Q. Chen, J. Wang, B. Li, and Y. Feng, "Synthesis and surface activities of amidobetaine surfactants with ultra-long unsaturated hydrophobic chains," Journal of Surfactants and Detergents, vol. 15, no. 5, pp. 657–661, 2012.

19. Z. Chu and Y. Feng, "Empirical correlations between Krafft temperature and tail length for amidosulfobetaine surfactants in the presence of inorganic salt," Langmuir, vol. 28, no. 2, pp. 1175–1181, 2012.

20. Y. Han, Y. Feng, H. Sun, Z. Li, Y. Han, and H. Wang, "Wormlike micelles formed by sodium erucate in the presence of a tetraalkylammonium hydrotrope," The Journal of Physical Chemistry B, vol. 115, no. 21, pp. 6893–6902, 2011.

21. Y. Han, Z. Chu, H. Sun, Z. Li, and Y. Feng, "'Green' anionic wormlike micelles induced by choline," RSC Advances, vol. 2, no. 8, pp. 3396–3402, 2012.

22. J. Wang, J. Wang, B. Wang, S. Guo, and Y. Feng, "Synthesis and aqueous solution properties of polyoxyethylene surfactants with ultra-long unsaturated hydrophobic chains," Journal of Dispersion Science and Technology, vol. 34, no. 4, pp. 504–510, 2013.

23. H. Wang, X. Cao, J. Zhang, and A. Zhang, "Development and application of dilute surfactant-polymer flooding system for Shengli oilfield," Journal of Petroleum Science and Engineering, vol. 65, no. 1-2, pp. 45–50, 2009.

24. H. K. Kotlar, O. Selle, and O. Torsaeter, "Enhanced oil recovery by COMB-flow: polymer floods revitalized," in Proceedings of

the SPE International Symposium on Oilfield Chemistry, Paper SPE 106421, pp. 540–545, Houston, Tex, USA, February-March 2007.

25. J. L. Cayias, M. E. Hayes, R. S. Schechter, and W. H. Wade, "Surfactant aging—possible detriment to the tertiary oil recovery," Journal of Petroleum Technology, vol. 28, pp. 985–988, 1976.

Effect of Polymer Adsorption on Permeability Reduction in Enhanced Oil Recovery

Saurabh Mishra, Achinta Bera, and Ajay Mandal

Department of Petroleum Engineering, Indian School of Mines, Dhanbad 826004, India

ABSTRACT

In order to reduce the permeability to water or brine, there is a possibility of polymer injection into the reservoir. In the present work, special focus has been paid in polymer [partially hydrolyzed polyacrylamide (PHPA)] injection as a part of chemical method. Tests were conducted in the laboratory at the ambient temperature to examine the reduction in permeability to water or brine in the well-prepared sand packed after the polymer injection. The experiments were performed to study the effect of polymer adsorption on permeability reduction by analyzing residual resistance factor values with different concentrations of polymer solutions. The rheological behavior of the polymer has also been examined. The experimental results also indicate that the adsorption behavior of polymer is strongly affected by salinity, solution pH, and polymer concentration. To investigate the effect of polymer adsorption and mobility control on additional oil recovery, polymer flooding experiments were conducted with different polymer concentrations. It has been obtained that with the increase in polymer concentrations, oil recovery increases.

INTRODUCTION

With reservoir getting matured, the increased water production with producing oil is a major concern in the petroleum industry. Hydrocarbon production decreases, which affects recovery economics and disposal of the excessive high amount of underground water, which causes complex environmental problems [1]. Oil and gas reservoirs are often heterogeneous, having a different permeability in different-different layers. This causes channelling of excessive water production through high permeability layers, as a result of which large amount of movable oil and gas remains trapped in low permeability zones which results in poor recovery in primary and secondary stages of production. A significant part of the residual oil can be recovered by application of a polymer solution in the heterogeneous reservoirs [2]. The excess water production can be controlled without affecting oil production rate by using polymer to reduce the relative permeability to water more than the relative permeability to oil [3]. The injection of polymer solutions in production wells has proven to be an effective method to reduce or block excessive water production [4]. In a water-wet reservoir, oil flows inside pores while water through the annulus between the pore walls and the oil-water interface [5]. When the polymer solution is injected, it gets adsorbed on the rock surface to form a thin layer. As water flows through this zone, the adsorbed polymer swells and prevents the flow of water while allowing the oil to flow through. When the polymer is injected, it builds up a resistance to flow in the reservoir through permeability reduction. This increased flow resistance diverts the injected water into unswept areas [6]. Interestingly, the adsorbed polymer is sustained over a long time; this leads to a great reduction in permeability [7]. In 1974, some of the basic concerns pertaining polymer adsorption in porous medium and effect of polymer adsorption on flow properties have been investigated by Hirasaki and Pope [8]. They reported that polymer treatment reduces permeability by adsorption of polymer and reduction in pore radii. The permeability reduction caused by polymer adsorption was traditionally known as "residual resistance factor" which is tantamount to the endpoint relative permeability [9].

Laboratory experiments have been performed by many researchers to examine polymer retention and its effect on oil recovery. Zaitoun and Bertin showed in their work that oil and water permeability can be modified by wall effects due to adsorption of polymer [10]. Grattoni et al. reported atomic force microscopy study on polymer retention on microglass surface. They discussed the basic adsorption mechanism and flow of polyacrylamide in the formation of layer thickness [11]. Lai et al. have recently showed the permeability reduction performance of a hyperbranched polymer in high permeability porous media [12]. Al-Hashmi and Luckham studied the effect of salt concentration on the adsorption of high-molecular-weight nonionic and cationic polymers on glass surface and showed that with increase in salt concentration the adsorbed layer thickness also increases [13]. Al-Sharji et al. reported that polymer adsorption significantly influences the water permeability and enhances the oil recovery [14].

In polymer flooding technique, a high-molecular-weight water-soluble polymer is added as a thickening agent into injected water to increase viscosity of aqueous phase for mobility control which can improve sweep efficiency. One of the attractive properties of such polymers is their ability to reduce the relative permeability to water more than the relative permeability to oil in porous media. Liu et al. reported in their framework in the Daqing oil field that from polymer flooding water cut was significantly reduced and oil recovery was improved by 10% [15]. Polymer flooding projects yielded additional recoveries of 10–15% of the original oil in place. Since the efficiency of polymer flooding differs with the rock and fluid properties as it varies from field to field. Thus an efficient work is required for improvement of this technology in practical field [16].

In this work, Partially Hydrolyzed Polyacrylamide (PHPA) is used to examine the effect of polymer adsorption on the permeability modification and oil recovery in a sand pack system. Rheological property of the used polymer has also been investigated by measuring the viscosity of different concentrated polymer solutions with a variation of shear rate. Adsorption experiments have been carried out with a variation of polymer concentrations, salinity, and pH of the solution. Along with this, thickness of the adsorbed layer of polymer has also been calculated with the help of the permeability reduction data. Flooding experiments have been performed to investigate the additional oil recovery after conventional water flooding.

EXPERIMENTAL SECTIONS

Materials Used

Partially hydrolyzed Polyacrylamide (PHPA) (Polymer Pusher 1000, SNF Floerger, France) with 98% purity was supplied by Central Drug House (CDH) Pvt. Ltd., India. Typical degree of hydrolysis of the used polymer is 30% to 35%. Sodium Chloride (NaCl) with 98% purity, procured from Qualigens Fine Chemicals, India, was used for preparation of brine. Reverse osmosis water from Millipore water system (Millipore SA, 67120 machine, France) was used for the preparation of solutions. Sand particles (60–70 mesh sized) were used as adsorbent in the adsorption study. The details compositional analysis of the used sand particles has been given in our previous work [17]. The crude oil used in the flooding experiments was collected from the Ahmedabad oil field (India). It was degassed and dehydrated before use. The physical properties of crude oil are shown as follows:

- viscosity: $5.12\,Pa{\cdot}s$ at $45°C$ at $20\,s^{-1}$ shear rate,
- total acid number: $0.038\,mg\,KOH/g$,
- API gravity: 38.5,
- color: blackish.

Experimental Procedures

Preparation and Rheology of Polymer Solution

Polymer solutions were prepared by dissolving the required amount of polymer in brine. Polymer solutions were prepared by the slow addition of the polymer powder to the brine in a vortex created by a magnetic stirrer. Gentle agitation was maintained over day for the complete polymer dissolution to obtain a homogeneous solution. Care was taken to rotate the magnetic stirrer at low rpm speed to avoid any mechanical degradation. In the present study, the rheological property of the used polymer was measured with the help of a rheometer (Physica MC1) at different concentrations. The flow curves were recorded between 10^{-1} and 10^3 1/s shear rate using a linear rate program and 2 min ramp time.

Permeability Reduction by Polymer Flooding

Sand packed was mounted horizontally in a sand packed holder and flow experiment was performed. First sand packed was flooded with brine and initial permeability to brine was calculated. Polymer solution was then injected, when sand packed was fully saturated with brine. The flow rate was measured during polymer flooding. When the flow rate attained a constant value, the polymer injection was stopped. The sand packed was then kept at rest for 50–56 hours for maximum interaction between sand particles and polymer molecules. Brine was then injected at constant pressure and flow rate curve was drawn. Above procedure was repeated for each concentration of polymer solution.

Polymer Adsorption and Thickness of Adsorbed Layers

A series of batch experiments were carried out to determine the adsorption of PHPA on the sand particles. 100 g of clean sand particles were added to 100 mL polymer solution in a 250 mL conical flux and allowed to conduct the experiment by constant shaking at 303 K for 24 h on a horizontal shaker mode. Core flooding is a dynamic process.

So the dynamic condition was considered to maintain the polymer flooding process. During the polymer injection pressure was applied therefore the dynamic condition is maintained. Actually dynamic condition is an appropriate field relevant condition. It is important to mention that the ratio of adsorbent to adsorbate also varies for different studies [18, 19]. In the present study, as flooding experiment was conducted on sand packed, the ratio was taken in this pattern. Again a certain amount of adsorbent is required for maximum adsorption. So a large excess concentration of adsorbent was taken in this study. After adsorption, the polymer solutions were isolated by centrifugation. The equilibrium or residual concentrations of the polymer solutions were determined by UV-spectrophotometer. The equilibrium concentrations (C_e) were calculated by using Beer-Lambert equation:

$$A = \varepsilon * L * C, \tag{1}$$

where A i is absorbent, ε is molar extinction coefficient (L/mole/cm), L is the path length (10 mm), and C is concentration (mole/L) of the solution.

The adsorption capacity of polymer on the adsorbent, Γi (mg/g), was calculated by a mass balance relation [20]:

$$\Gamma i = (C_0 - C_e)\frac{V}{m}, \tag{2}$$

where C_0 and C_e are the initial and equilibrium concentrations of polymer solutions (mg/g), respectively, V is the volume of the polymer solution (L), and m is the weight of the sand particles (g) (adsorbent) used. The effect of salt concentration, pH, and polymer concentrations is investigated. These may greatly affect polymer adsorption hence permeability reduction and polymer flood efficiency.

The thickness of the adsorbed layer of polymer (χ_p) can be calculated from the permeability reduction data. The calculation is based on the assumption of Poiseuille's fluid flow through a capillary having its cross-section reduced by a uniform layer. The thickness of the adsorbed layer of polymer is calculated as follows [3]

$$\chi_p = r_p (1 - RRF^{-1/4}), \tag{3}$$

where RRF is the residual resistance factor and is the average pore diameter, which can be calculated as [3]

$$rp = 1.15 \left(\frac{8k_w}{\varphi} \right)^{1/2}, \qquad (4)$$

where k_w is the effective permeability to water and porosity of the sand pack.

Experimental Apparatus and Method for Polymer Flooding

The experimental apparatus is composed of a sand pack holder, cylinders for holding chemical slugs and crude oil, positive displacement pump, and measuring cylinders for collecting the samples. The detail of the apparatus and method of flooding is reported in our earlier paper [21]. The schematic diagram of flooding experiment is given in Figure 1. The model geometry size is cm and cm. The sand packed holder was tightly packed with uniform sands (60–100 mesh) and saturated with 1.0wt% brine. The additional recoveries were calculated by material balance.

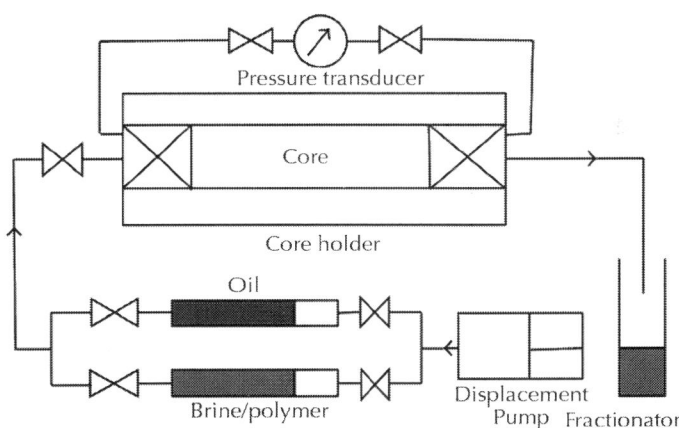

Figure 1: Schematic diagram of flooding experiment.

Darcy's law was used to calculate the effective permeability to oil (k_o) and effective permeability to water (k_w) at irreducible water saturation (S_{wi}) and residual oil saturation (S_{or}), respectively. For a horizontal linear system, flow rate is related to permeability as follows:

$$q = \frac{kA}{\mu}\frac{dp}{dx},$$ (5)

where q is the volumetric flow rate (cm^3 /Sec.), A is the total cross-sectional area of the sand pack (cm^2), μ is the fluid viscosity (cp), dp/dx is the pressure gradient (atm/cm), and k is permeability in Darcy. Experiments were performed using 2500 ppm and 2000 ppm PHPA solutions.

RESULTS AND DISCUSSION

Rheology of PHPA Solutions

Polymer rheology plays an important role in controlling the mobility ratio and hence the sweep efficiencies. The basic rheological behavior of aqueous solutions of partially hydrolyzed polyacrylamide has been investigated by varying polymer concentrations. Figure 2 shows the variation of apparent viscosity with shear rate. The viscosity of solutions increased with increasing polymer concentrations because of the increasing intermolecular entanglement. The viscosities of all samples decreased with the increasing shear rate, suggesting that the aqueous solution of PHPA exhibits non-Newtonian behavior. This is due to uncoiling and aligning of polymer chains when exposed to shear flow. As the shear rate approaches toward zero the solutions show maximum viscosity. The viscosity is quite significant at moderate shear rate. It has also been found that at higher shear rate the solution approaches towards Newtonian behavior.

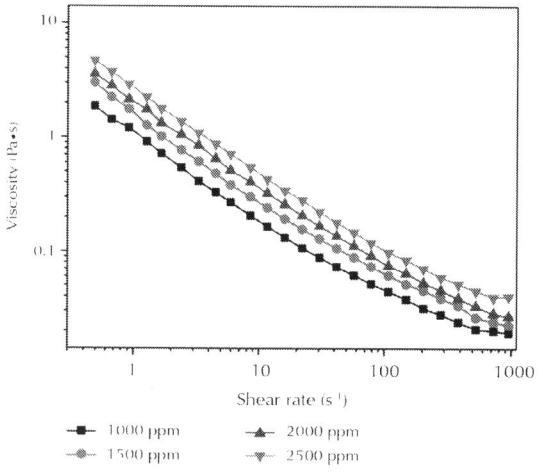

Figure 2: Viscosity of PHPA solutions with different concentrations at 303 K.

Permeability Reduction

The flow behavior of polymer solutions of different concentrations and subsequent brine injection through the sand packs are shown in Figure 3. It may be found that as the cumulative injected volume of polymer solution increases the flow rate gradually decreases. This is because of adsorption of polymer on the sand particle. After a certain pore volume injected there is no further change in flow rate. This indicates that the surfaces of sand particles are now fully saturated with polymer and there is no further adsorption. The system is kept as it is for around 50 hours for maximum interaction between polymer and sand particles. The polymer coats on the rock surfaces with a hydrophilic film. As water passes over the polymer, the film swells, reducing the effective permeability. In presence of oil, the swelling does not occur, thereby improving potential oil recovery [22]. Then the sand packs are continuously injected with brine. Brine flow rate gradually increases due to desorption of polymer. But after a certain pore volume of brine injected it has been found that there is no further increase of flow rate. This suggests that even after a long time of chase water flooding after polymer flooding, a certain amount of polymer is retained and hence the benefits of permeability reduction can be achieved for a long time.

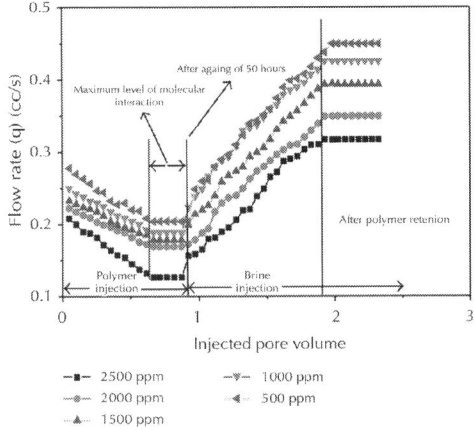

Figure 3: Flow rate curves during injection of brine after aging of polymer on sand grains.

The absolute permeability and the residual permeabilities have been measured by Darcy's law and the results are shown in Figure 4. It has been found from the figure that absolute permeability of the sand pack is much higher than that of residual permeabilities. And the residual permeabilities increase with the increase in concentrations of polymer.

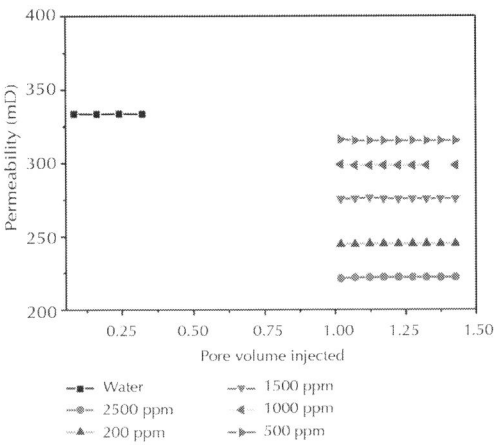

Figure 4: Reduction in permeability due to polymer retention.

To examine restriction in flow due to polymer adsorption, resistance factor (RF) and residual resistance factor (RRF) ware calculated. Figure 5 shows the values of RF and RRF with a variation of pore volume injected at different polymer concentrations. Initially, when the polymer was injected, it prefers high permeability zones due to low resistance in the flow. Results show that the increment in the polymer concentrations brings to the high value of RF and RRF. Interaction between polymer and sand grains leads to profile modification. This profile modification depends on how polymer is adsorbed on the wall of sand grains in high permeability zones. Due to polymer adsorption pore throat radii reduces, this results in an unexpected decrement in mobility to polymer. The measure of the mobility reduction is known as "resistance factor." Resistance factor was calculated by the following equation:

$$RF = \frac{\lambda_w}{\lambda_p},$$ (6)

where λ_w is the mobility of water and λ_p is the mobility of polymer solution.

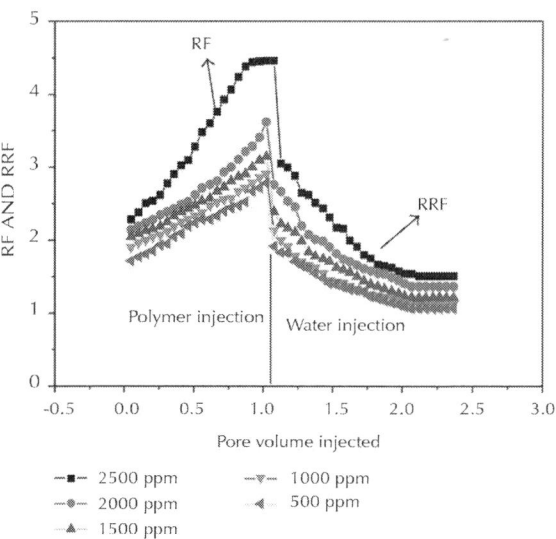

Figure 5: Plot of RF and RRF with different concentrations of polymer solution.

Mobility is calculated by using Darcy's equation as follows:

$$\lambda = \frac{q * L}{A * \phi * \Delta p}$$

(7)

Table 1 shows the calculated data of RF and RRF for PHPA with different concentrations.

Table 1: Calculated values of RF and RRF for different concentrated PHPA solutions

Test samples	Mobility	Resistance factor (RF)	Residual resistance factor (RRF)
Water	1.027	—	—
C_1 (2500 ppm + 2 wt% NaCl)	0.231	4.461	1.504
C_2 (2000 ppm + 2 wt% NaCl)	0.284	3.621	1.361
C_3 (1500 ppm + 2 wt% NaCl)	0.327	3.143	1.211
4_1 (1000 ppm + 2 wt% NaCl)	0.353	2.905	1.119
C_5 (500 ppm + 2 wt% NaCl)	0.369	2.783	1.059

Polymer Adsorption

Adsorption of polymer on rock surfaces is influenced by several factors. The influencing factors are polymer concentrations, salinity of the solution, the pH of the solution, temperature, and so forth. Some of the factors have been discussed here with experimental results.

Thickness of Adsorbed Layer

Figure 5 shows the relation between polymer concentrations and thickness of adsorbed layer. It is confirmed from Figure 6 that with the increase in polymer concentrations the thickness of adsorbed layer also increases. As concentration increases, the number of molecules increases, thereby the probability of interaction between polymer molecules and sand particles increases. This brought increment in adsorption capacity per unit area. Therefore, it is confirmed that the adsorption layer thickness increases with an increase in adsorption. The increase in the adsorption layer thickness as the polymer content in solution increases is due to a change in the conformation of the adsorbed molecules [23]. The change in the adsorption layer thickness in the initial segments of the isotherms suggests that at low concentrations of polymer, the molecules on the surface of the sand particles are distributed and bound with a small number of segments. The increase in concentrations of polymer causes a rearrangement of the structure of the adsorbed layer. The adsorbed molecules break previously formed bonds so that they straighten and unfold and the layer thickness increases.

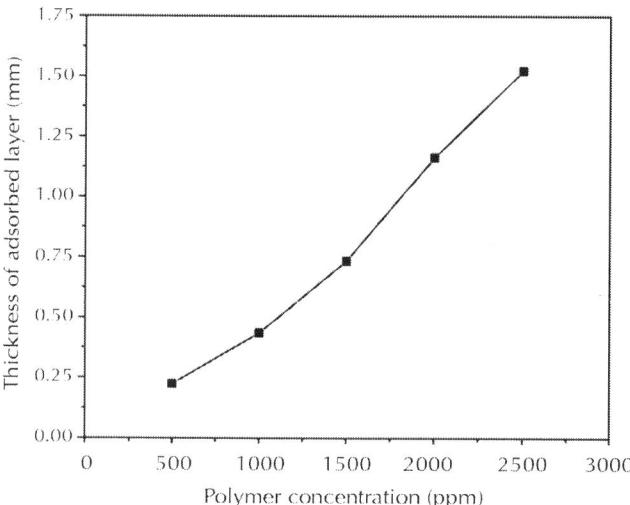

Figure 6: Relation between polymer concentrations and thickness of adsorbed layer.

Effect of Polymer Concentrations on Adsorption

Figure 7 shows the time dependent polymer adsorption on the sand surface at different concentrations. With increase in time polymer adsorption increases and after certain time adsorption remains constant with time. As time increases, the number of adsorption sites decreases due to formation of adsorbed polymer layer on the sand surface. When all the sites are covered with polymer, then further adsorption does not take place and time independent adsorption occurs. Figure 7 shows that with the increase in polymer concentrations the polymer adsorption capacity also increases. At higher concentrations of polymer solutions, number of polymer molecules increases, which in turn increase the probability of interaction between the sand surface and polymer molecule. It is also important to point out here that after certain concentration no further adsorption takes place due to saturation of the adsorption capacity of the active adsorption sites [9, 24, 25]. In the present case, only four different concentrations have been used to show the effect of concentrations on polymer adsorption onto the sand surface and it has been found that up to 2500 ppm concentration of PHPA polymer adsorption increases. Adsorption causes the permeability reduction. Ogunberu and Asghari have reported on their experimental work that increase in polymer concentrations results in an increase in permeability reduction caused by polymer adsorption and efficiency of polymer flooding [3].

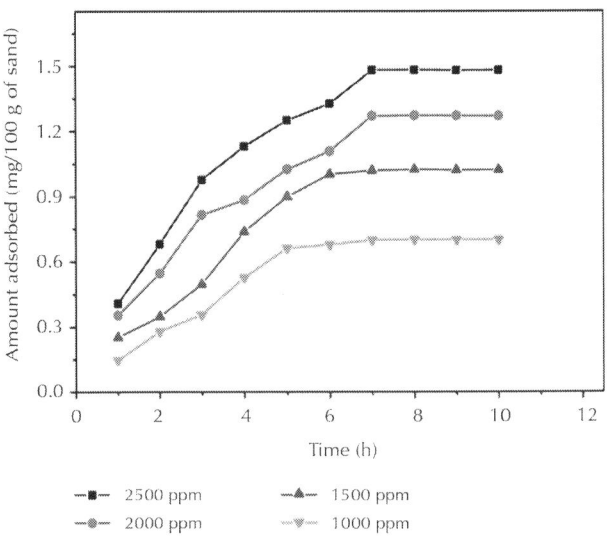

Figure 7: Effect of polymer concentrations on polymer adsorption.

Effect of Solution pH on Polymer Adsorption

The adsorption behavior of PHPA at different solution pH has been shown in Figure 8. The adsorption capacity increases when solution pH is decreased as shown in Figure 8. At low pH, sand surfaces become more positive due to the acidic nature of the polymer solutions and the interaction of sand surface with anionic polymer PHPA is high, hence, adsorption capacity is high. The negatively charged carboxyl group dissociated through electrostatic attraction to the positively charged ions (due to more H^+ ions). This causes more attraction between polymer molecules and rock surfaces. When the pH of the solution increases, that is, the solutions become alkaline in nature, the sand surfaces become more negative. The repulsion between PHPA with negatively charged carboxylic acid group and negatively charged sand surface takes place, which leads to lower adsorption of the polymer on sand surfaces. So, it is confirmed from the study of the effect of pH on adsorption that the adsorption of polymer on rock surfaces can be modified by fixing the solution pH which is important for polymer flooding.

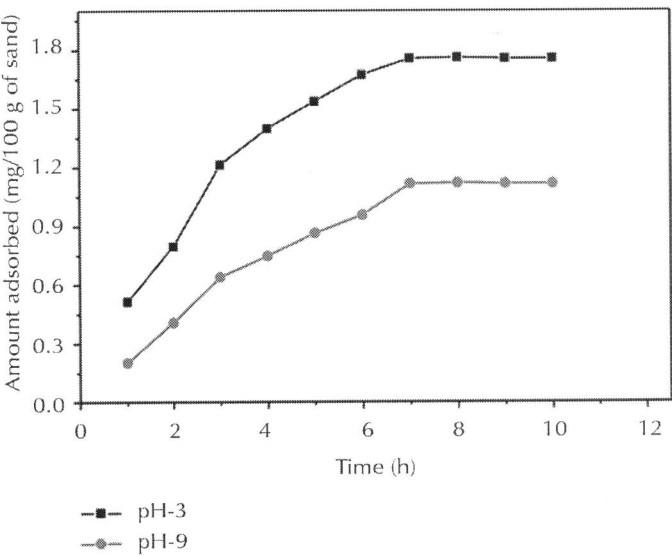

Figure 8: Effect of solution pH on polymer adsorption on sand surface.

Effect of Salt Concentration on Polymer Adsorption

Experimental results have been shown in Figure 9. It is clear from Figure 8 that polymer adsorption increases with increase in salinity. As the salt concentration increases, charge shielding takes place due to positively charged ions of the salt (Na^+), thus the hydrodynamic radius of polymer molecule reduces. Due to this intermolecular interaction, electrostatic repulsion in the polymer solution decreases. This causes an increase in adsorption capacity of polymer capacity of polymer solution. Moreover the salt concentration may affect the displacement efficiency and volumetric sweep efficiency of the polymer flood. High salt concentration may affect the polymer absorption by an increase in the viscosity of polymer due to electrostatic repulsion in the polymer solution and a reduction in the surface area of adsorbent, access to polymer molecules. Due to the above two parametrical effects, adsorption capacity increases as salt concentration increases.

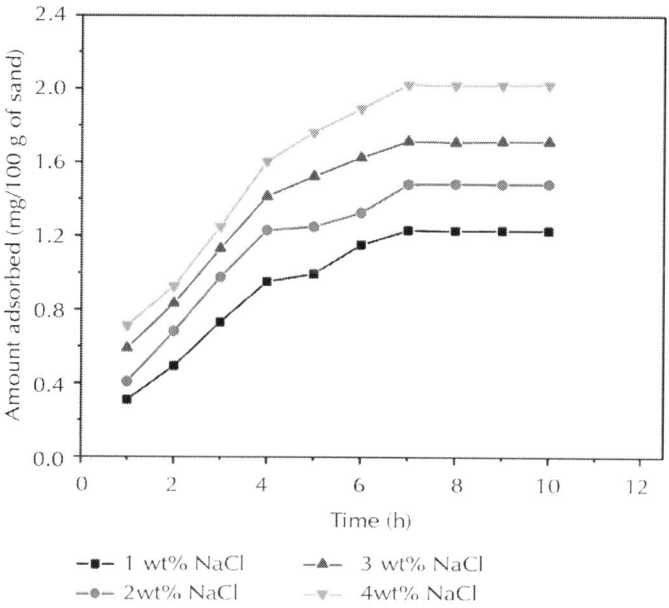

Figure 9: Effect of salt concentration on polymer adsorption.

Oil Recovery by Polymer Flooding

Injection of the polymer increases the recovery of oil by improving the mobility ratio, reducing the effective permeability to water and increasing the effective permeability to water. In the present study, two different concentrations of PHPA were used to measure the additional recovery of oil after conventional water flooding. Initially the sand pack was flooded with brine. When water cut reached around 95%, approximately 0.75 PV polymer solution was injected to investigate its effect on the efficiency of polymer flooding. Two sets of experiments were performed using 2000 ppm and 2500 ppm polymer solution. The recovery of oil while flooding the sand packed with brine and different concentrations of PHPA are shown in Figure 10. The main parameters and results of these flood tests are in Table 2, including the initial water saturations, the injected pore volumes and oil recoveries during water flooding and tertiary recoveries. All recoveries are calculated based on the original oil in place (OOIP).

Table 2: Porous media property and oil recovery by polymer flooding

Experiment number	Porosity (%)	Permeability, k(mD)		Design of polymer slug for flooding	Recovery of oil by water flooding at 95% water cut (% OOIP)	Additional recovery (% OOIP)	% Saturation		
		$k_w(S_w=1)$	$k_0(S_{wi})$				S_{wi}	S_{oi}	S_{or}
P1	32.5	333.5	58.33	0.75 PV polymer (2500 ppm) + chase water	43.9	21.00	12	88	23.5
P2	32.4	333.6	58.4	0.75 PV polymer (2000 ppm) + chase water	44.0	19.21	12	88	23.8

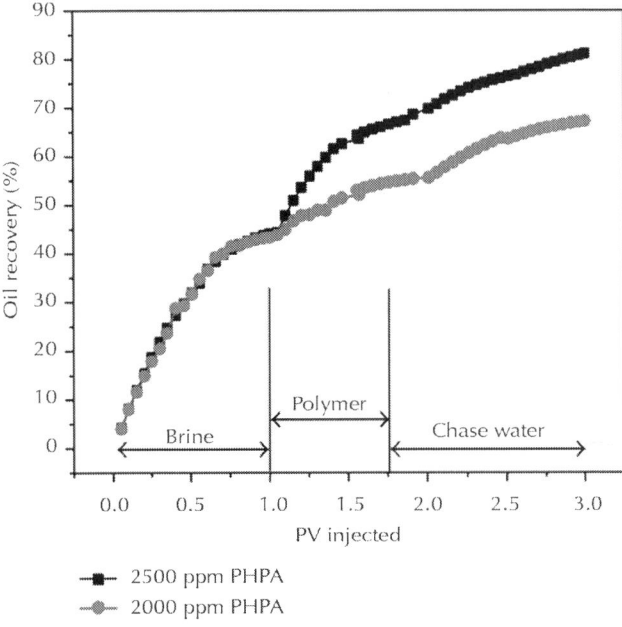

Figure 10: Oil recovery by polymer flooding using 2500 ppm and 2000 ppm PHPA solutions.

The additional oil recoveries by polymer flooding were 19.21% and 21.00% OOIP for 2000 ppm and 2500 ppm PHPA solutions. These results were compared with another literature reported by Maitin, where the laboratory investigations, simulation studies, and a pilot test in the oil fields of a North German sediment basin have produced the incremental recovery of 8 to 22% OOIP [26].

CONCLUSIONS

Rheological properties of the PHPA solution have been determined and it has been found that with increase in polymer concentrations viscosity also increases. Adsorption of polymer on sand particles has been studied by polymer flooding through sand packed. Experimental results show that after aging, there was significant interaction between polymer molecules and sand grains, which plugged the flow path to inject brine at constant pressure; as a result, there was a significant decrement in flow rate. As brine injection increased, polymer retention decreased and reached a critical value which gives an absolute permeability reduction. The result shows that increase in polymer concentrations enhances the polymer adsorption capacity. At low pH, PHPA adsorption capacity on sand surface is high due to the acidic nature of the solution which makes the sand surface more positive and that is why the interaction of sand surface with anionic polymer PHPA is high, hence, adsorption capacity is high. When the pH of the solution increases the solution becomes alkaline and as a result repulsion between PHPA and negatively charged sand surface increases, therefore, adsorption of the polymer decreases. As the salt concentration increases, charge shielding takes place due to positively charged ions of the salt (Na^+), thus the hydrodynamic radius of polymer molecule reduces. Due to this intermolecular interaction, electrostatic repulsion in the polymer solution decreases. This causes an increase in adsorption capacity of polymer capacity of polymer solution. The result shows that with the increase in polymer concentrations, thickness of adsorbed layer also increases. To investigate the relationship between the incremental tertiary oil recovery and the effective viscosity of polymer solutions, two sets of sand packed flood tests were carried out by injecting polymer solutions of different viscosities. The additional oil recoveries by polymer flooding were 19.21% and 21.00% OOIP for 2000 ppm and 2500 ppm PHPA solutions, respectively.

ACKNOWLEDGMENTS

The authors gratefully acknowledge the financial assistance provided by University Grant Commission [F. no. 37-203/2009 (SR)], New Delhi, to the Department of Petroleum Engineering, Indian School of Mines, Dhanbad, India. Thanks are also extended to all individuals associated with the project.

REFERENCES

1.   P. L. J. Zitha, K. G. S. van Os, and K. F. J. Denys, "Adsorption of linear flexible polymers during laminar flow through porous media: effect of concentration," in Proceedings of the SPE/DOE

Improved Oil Recovery Symposium, pp. 19–22, Tulsa, Okla, USA, April 1998, SPE paper no. 39675.

2.   O. L. Kouznetsov, E. M. Simkin, G. V. Chilingar, and S. A. Katz, "Improved oil recovery by application of vibro-energy to waterflooded sandstones," Journal of Petroleum Science and Engineering, vol. 19, no. 3-4, pp. 191–200, 1998.

3.   A. L. Ogunberu and K. Asghari, "Water permeability reduction under flow-induced polymer adsorption," in Proceedings of the 5th Canadian International Conference, Alberta, Canada, June 2004, paper no. 2004-236.

4.   P. D. Moffitt, "Long-term production results of polymer treatments in producing wells in western Kansas," Journal of Petroleum Technology, vol. 45, no. 4, pp. 356–362, 1993

5.   A. Zaitoun and N. Kohler, "Two phase flow through porous Media: effect of an Adsorbed Polymer Layer," in Proceedings of the SPE Annual Technical Conference and Exhibition, Huston, Tex, USA, October 1998, SPE paper no. 18085.

6.   M. A. Mohammed, "Investigation of polymer adsorption on rock surface of high saline reservoirs," inProceedings of the Saudi Arabia Section Technical Symposium (SPE '08), Alkhobar, Saudi Arebia, May 2008, SPE paper no. 120807.

7.   H. T. Dovan and R. D. Hutchins, "New polymer technology for water control in gas wells," SPE Production and Facilities, vol. 9, no. 4, pp. 280–286, 1994.

8.   G. J. Hirasaki and G. A. Pope, "Analysis of factors influencing mobility and adsorption in the flow of polymer solution through porous media," Society of Petroleum Engineering Journal, vol. 14, no. 4, pp. 337–346, 1974

9.   C. G. Zheng, B. L. Gall, H. W. Gao, A. E. Miller, and R. S. Bryant, "Effects of polymer adsorption and flow behavior on two-phase flow in porous media," SPE Reservoir Evaluation & Engineering, vol. 3, no. 3, pp. 216–223, 2000.

10.   A. Zaitoun and H. Bertin, "Two-phase flow property modification by polymer adsorption," inProceedings of the SPE Improved Oil Recovery Symposium, Tulsa, Okla, USA, 1998, SPE Paper no. 39631.

11. B. C. A. Grattoni, P. F. Luckham, X. D. Jing, L. Norman, and R. W. Zimmerman, "Polymers as relative permeability modifiers: adsorption and the dynamic formation of thick polyacrylamide layers," Journal of Petroleum Science and Engineering, vol. 45, no. 3-4, pp. 233–245, 2004.

12. N. Lai, X. Qin, Z. Ye, C. Li, K. Chen, and Y. Zhang, "The study on permeability reduction performance of a hyperbranched polymer in high permeability porous medium," Journal of Petroleum Science and Engineering, vol. 112, pp. 198–205, 2013.

13. A. R. Al-Hashmi and P. F. Luckham, "Characterization of the adsorption of high molecular weight non-ionic and cationic polyacrylamide on glass from aqueous solutions using modified atomic force microscopy," Colloids and Surfaces A, vol. 358, no. 1-3, pp. 142–148, 2010

14. H. H. Al-Sharji, C. A. Grattoni, R. A. Dawe, and R. W. Zimmerman, "Disproportionate permeability reduction due to polymer adsorptrion-entanglement," in SPE European Formation Damage Conference, The Hague, Netherlands, 2001, SPE Paper no. 68972.

15. S. Liu, D. L. Zhang, W. Yan, M. Puerto, G. J. Hirasaki, and C. A. Miller, "Favorable attributes alkaline-surfactant-polymer flooding," Society of Petroleum Engineering Journal, vol. 13, no. 1, pp. 5–16, 2008.

16. Y. Masuda, K.-C. Tang, M. Miyazawa, and S. Tanaka, "1D simulation of polymer flooding including the viscoelastic effect of polymer solution," SPE Reservoir Engineering, vol. 7, no. 2, pp. 247–252, 1992.

17. A. Bera, T. Kumar, K. Ojha, and A. Mandal, "Adsorption of surfactants on sand surface in enhanced oil recovery: isotherms, kinetics and thermodynamic studies," Applied Surface Science, vol. 284, pp. 87–99, 2013.

18. W. Lv, B. Bazin, D. Ma, Q. Liu, D. Han, and K. Wu, "Static and dynamic adsorption of anionic and amphoteric surfactants with and without the presence of alkali," Journal of Petroleum Science and Engineering, vol. 77, no. 2, pp. 209–218, 2011.

19. M. Safian-Boldani, M. P. Shahri, M. Zargartalebi, and M. Arabloo, "New surfactant extracted from zizyphus spina christi for enhanced oil recovery: experimental determination of static

adsorption isotherm," Journal of the Japan Petroleum Institute, vol. 56, no. 3, pp. 142–149, 2013.

20. M. A. Ahmadi and S. R. Shadizadeh, "Experimental investigation of adsorption of a new nonionic surfactant on carbonate minerals," Fuel, vol. 104, no. 2, pp. 462–467, 2013.

21. A. Samanta, K. Ojha, A. Sarkar, and A. Mandal, "Mobility control and enhanced oil recovery using partially hydrolyzed polyacrylamide (PHPA)," International Journal of Oil, Gas Coal Technology, vol. 6, no. 3, pp. 245–258, 2013.

22. D. D. Sparlin, "Evaluation of polyacrylamides for reducing water production (includes associated papers 6561 and 6562)," Journal of Petroleum Technology, vol. 28, no. 5, pp. 906–914, 1976

23. Y. Cohen and A. B. Metzner, "Adsorption effects in the flow of polymer solutions through capillaries,"Macromolecules, vol. 15, no. 5, pp. 1425–1429, 1982.

24. N. Tekin, A. Dinçer, Ö. Demirbaş, and M. Alkan, "Adsorption of cationic polyacrylamide (C-PAM) on expanded perlite," Applied Clay Science, vol. 50, no. 1, pp. 125–129, 2010.

25. N. Tekin, Ö. Demirbaş, and M. Alkan, "Adsorption of cationic polyacrylamide onto kaolinite,"Microporous and Mesoporous Materials, vol. 85, no. 3, pp. 340–350, 2005

26. B. K. Maitin, "Performance analysis of several polyacrylamide floods in the North German oil fields," inProceedings of the SPE/ DOE Enhance Oil Recovery Symposium, Tulsa, Okla, USA, April 1992, SPE paper no. 24118.

A Review of CO$_2$ Sequestration Projects and Application in China

Yong Tang, Ruizhi Yang, and Xiaoqiang Bian

The State Key Laboratory of Oil & Gas Reservoir Geology and Exploitation Engineering, Southwest Petroleum University, Chengdu 610500, China

ABSTRACT

In 2008, the top CO$_2$ emitters were China, United States, and European Union. The rapid growing economy and the heavy reliance on coal in China give rise to the continued growth of CO$_2$ emission, deterioration of anthropogenic climate change, and urgent need of new technologies. Carbon Capture and sequestration is one of the effective ways to provide reduction of CO$_2$ emission and mitigation of pollution.

Coal-fired power plants are the focus of CO_2 source supply due to their excessive emission and the energy structure in China. And over 80% of the large CO_2 sources are located nearby storage reservoirs. In China, the CO_2 storage potential capacity is of about 3.6×10^9 t for all onshore oilfields; 30.483×10^9 t for major gas fields between 900 m and 3500 m of depth; 143.505×10^9 t for saline aquifers; and 142.67×10^9 t for coal beds. On the other hand, planation, soil carbon sequestration, and CH_4–CO_2 reforming also contribute a lot to carbon sequestration. This paper illustrates some main situations about CO_2 sequestration applications in China with the demonstration of several projects regarding different ways of storage. It is concluded that China possesses immense potential and promising future of CO_2 sequestration.

INTRODUCTION

The enormous emission from greenhouse gas, predominated by CO_2, has caused increasing threat to human environment and the ecological system. The current global annual carbon emission reaches up to more than 30 billion tons. In China, fossil fuel takes up 92.6% of the total energy; 67.1% of CO_2 is generated from coal and petroleum. Moreover, China is the biggest CO_2 emitter by now. According to International Environment Agency, emission from China would overpass the whole world's CO_2 emission by 2020 [1]. Therefore, it is an urgent requirement for China to transform from high-carbon to low-carbon society.

According to "Report on the Development of Low Carbon Economy of China (2012)," China is the largest country for carbon emission reduction. The world's largest carbon emission reduction project started in China in 2005, which is expected to reduce about 19 million tons of CO_2 equivalent emission every year [2]. 1.5 billion tons of CO_2 emission has been reduced during "11th five-year plan" in China, and it is likely to cut 7 billion tons of CO_2 in 2020.

Various ways of reducing carbon emission have already been applied in China. And, among them, a major mitigation method is carbon capture and sequestration (CCS).

It is believed that CCS is the long-term isolation of carbon dioxide from the atmosphere through physical, chemical, biological, or engineered process. It includes carbon sequestration through forestation, soil carbon sequestration, direct ocean injection of CO_2

either into the deep seafloor or into the intermediate depths, and the deep geological sequestration, or even direct conversion of CO$_2$ to carbonate minerals [3], of which geological sequestration is a major component. CCS is an effective way for China to alleviate pollution and enhance the oil recovery, and most underground spaces in China are good for CO$_2$ geological storage [4]. However, CCS has just started in China, and there is a certain gap between China and abroad. But there are still some technical foundations in China, especially in the area of CO$_2$ recycling and injection [5].

Several main types of geological storage media for carbon sequestration are mostly considered in China: depleted or active oil and natural gas field, coal layers, and deep saline aquifers. The win-win effects make oil and natural gas field and coal layers are the promising storage media with great advantages. By using CO$_2$ for oil and gas fields and the coal seams, CO$_2$ is stored and the production is increased. And the deep saline aquifers are attractive due to the large storage capacity of interest [6–9].

Figure 1 shows a map of large (100+ kt CO$_2$/yr) CO$_2$ sources and potential candidates for geologic CO$_2$ storage basins in China [10].

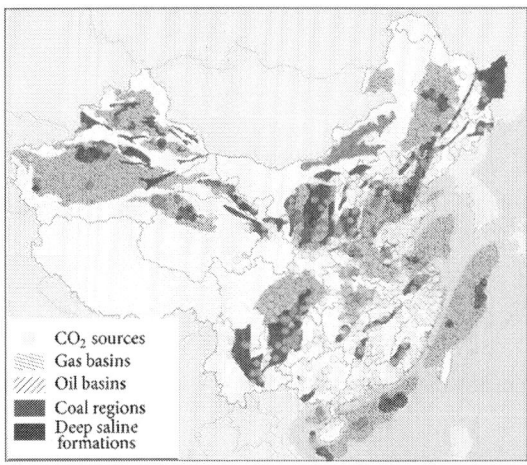

Figure 1: Locations of large CO$_2$ point sources and CO$_2$ storage reservoir in China (from Dahowski et al. [10]).

CO_2 SOURCE SUPPLY

A large amount of CO_2 emitted by industry could be supposed to serve as the significant potential CO_2 source to meet the storage demand if only the advanced capturing technology is available. And coal-fired power plant is the focus of CO_2 capture due to its excessive emission and the energy structure in China [3, 11, 12]. Therein, technologies of solvents method, membranes separation, solid sorbents, and cryogenic fractionation have been applied to separate CO_2 from natural gas or waste gas [13]. CO_2 could be transported via highway, railway, shipping, and pipeline, of which pipeline is especially suitable for large-scaled and long-term gas injection, like the CO_2-EOR project in Jilin oilfield.

Many efforts have been used to develop more efficient techniques for CO_2 capture in China, like the blended solvent presented by the Joint International Center for CO_2 Capture and Storage of Hunan University, MSA chemical absorption technique developed by Sinopec. And the research of Joint Research Center for Advanced Environmental Technology of Tsinghua University showed that carbon-based materials have high adsorption capacity with merits of low cost and easy regeneration. And Fang indicated that membrane vacuum regeneration has the potential to reduce energy consumption greatly [14].

CO_2 capturing projects have been progressing extraordinarily throughout China. Post-Combustion Capture CO_2 and Refining Utilization project with capacity of 0.12×10^6 t/yr in China is the biggest postcombustion capture project in the world then [15]. Sinopec has built the 100 t/d CCUS (Carbon Capture, Utilization and Storage) project on coal-fired power plant flue gas and deployed three ways of recycling CO_2 with more than 80% of capture efficiency and over 95% of purity. China Huaneng Group has built the first coal-fired power plant CO_2 capture demonstration project in 2008 with 3000 t/yr of CO_2 capture ability and completed the second power plant in Shanghai Shidongkou demonstration project with 0.1×10^6 t/yr of CO_2 capture ability. Shenhua Group launched China's first CO_2 capture and geologic storage full process demonstration project in 2010 [16]. Moreover, the project with the scale of 50000 t/yr capture capacity which has product purity of more than 99.5% has been put into use in 2012 in Yanchang. And for the future, improving efficiency and reducing cost are the crucial development tendency.

CO$_2$ SEQUESTRATION

Estimation of CO$_2$ Sequestration Capacity

Several methods have been developed to assess the CO$_2$ storage capacity in geological media at home and abroad [17–27]. Examples are listed as follows.

Zhang et al. [18] developed the formula which considers the different storage mechanisms:

$$M_{CO_2} = M_1 + M_2 + M_3 + M_4, \qquad (1)$$

where M_1 is the storage capacity of CO$_2$ taking the volume previously occupied by produced oil; M_2 is the storage capacity of CO$_2$ dissolved in residual oil; M_3 is the storage capacity of CO$_2$ dissolved in water contained in reservoir; and M_4 is the storage capacity of CO$_2$ reacting with reservoir rock.

Sun and Chen [19] proposed the study to calculate increased oil production and the CO$_2$ storage capacity in oil reservoir and depleted oil reservoir.

Proportion of increased oil by CO$_2$-EOR is as follows:

$$\%EXTRA = \begin{cases} 5.3\% & (API \le 31) \\ (1.3 \times API - 35)\% & (31 < API < 41) \\ 18.3\% & (API \ge 41), \end{cases}$$

$$OOIP_e = OOIP \times C, \qquad (2)$$

Where OOIP is the original oil in place, Mt; C is the contact ratio between oil and CO$_2$. $OOIP_e$ is the amount of oil that can contact with CO$_2$, Mt.

The increased oil production and storage capacity are as follows:

$$EOR = OOIP_e \times \%EXTRA,$$

$$CO_2 = EOR \times R_{CO_2}, \qquad (3)$$

Where EOR is the increased oil production, Mt; CO_2 is the storage capacity, t or Mt; $R_{co,}$ is the ratio between the amount of injected CO_2 and the amount of increased oil, t/bbl or t/t.

However, for the CO_2 storage capacity in depleted oil reservoir,

$$CO_2 = OOIP \times RF\_O \times FVF\_O \times \rho CO_2, \tag{4}$$

Where RF_O is the oil recovery when depleted; FVF_O is the formation volume factor; ρCO_2 is the density of $SCCO_2$ under the reservoir temperature and pressure, Mt/m^3.

Tanaka and coworkers [20, 21] set up two models based on underground structures: model (5) is suitable for aquifers that are well sealed by cap rocks and model (6) for aquifers in monoclonal structures and there may be problem of CO_2 leakage into the upper portion. Consider

$$MCO_2 = Ef \times A \times h \times \Phi \times \rho$$
$$\times \left[\frac{Sg}{Bg(CO_2)} + (1 - Sg) Rs(CO_2) \right], \tag{5}$$

$$MCO_2 = Sf \times A \times h \times \Phi \times Rs(CO_2) \times \rho, \tag{6}$$

where Ef is the sweep efficiency (fraction, dimensionless), A is storage area (m^2), h is effective formation thickness (m), θ is effective reservoir porosity (fraction, dimensionless), Sg is saturation of supercritical CO_2 (fraction, dimensionless), $Bg(CO_2)$ is CO_2 formation volume factor (m^2/m^3, reservoir volume/standard volume), $RS(CO_2)$ is CO_2 solubility in formation water (m^3/m^2), ρ is density of CO_2 at standard condition (kg/m^3), and Sf is the storage factor (fraction, dimensionless).

Geological Sequestration

CO_2 can be more effectively sequestrated at pressure higher than 7.38 MPa (equivalent depth of about 800 m), and at temperature above

31.1°C, where CO$_2$ will stay in a supercritical state with an elevated density up to 600 kg/m^3, 400 times more condensed compared to that at atmospheric conditions. SCCO$_2$ (supercritical CO$_2$) is characterized by stable and inert chemical property. Consequently, at pressures and temperatures typically encountered in the field, CO$_2$ will behave as a supercritical fluid [28].

CO$_2$ geose questration has been implemented successfully around the world like CO$_2$-EOR and storage in Weyburn project of Canada in 2000 [29, 30]; CO$_2$ storage in K12-B gas field of The Netherlands in 2004 [31]; the upcoming ROAD project in 2015 with CO$_2$ storage in P18-4 depleted gas field of The Netherlands [32]; associated CO$_2$ separation and injection into the saline aquifer in Sleipner project of Norway in 1996 [33]; CO$_2$ storage in In Salah aquifer of Algeria in 2004 and Snohvit aquifer of Norway in 2008 [34, 35]; CO$_2$-enhanced coal bed methane (CO$_2$-ECBM) and storage in San Juan Basin of New Mexico in 1995 [36], and other CO$_2$-ECBM projects in USA [37, 38].

Research results suggest that CCS can provide a valuable greenhouse gas mitigation option for most regions and industrial sectors in China and can be able to store more than 80% of emissions from these large CO$_2$ sources (2900 million tons of CO$_2$ annually) at costs less than $70/t CO$_2$ for perhaps a century or more [10]. Similarly, various geose questration projects have been in progress in China, regarding the storage in oil and gas fields, in saline aquifer and in coal seams.

CO$_2$ Sequestration in Oil and Gas Field

Carbon sequestration with enhanced oil recovery (CSEOR) is a kind of win-win process to increase oil production and store CO$_2$. Moreover, the revenue created could be able to offset the storage cost and bring valuable profit.

CO$_2$ has been widely used for EOR around the world. CO$_2$-EOR projects now produce about 0.35 × 10^6 bbls/day in USA, accounting for 5.6% of total USA oil and gas production, compared to just 0.19 × 10^6 bbls/day in 2000. And approximately 50 million metric tons of CO$_2$ is used each year for EOR in USA [39, 40].

CSEOR or CSEGR has been assessed and applied for several oil and gas fields across China. When the buried depth is more than 800 m (guarantee the supercritical state of CO$_2$); the CO$_2$ storage potential capacity is of about 3.6×10^9 t, assuming that all onshore oilfields in China are used for CO_2-EOR, and it can reach up to 4.6×10^9 t while considering all onshore oilfields as depleted reservoirs. Therein, reservoirs in northeast and north China have tremendous sequestration potential, accounting for more than 60% of the total capacity [24].

Considering the depth between 900 m and 3500 m, China's major gas fields are able to provide storage capacity of about 30.483×10^9 t of CO_2, and the proven natural gas resources correspond to storage capacity of 4.103×10^9 t CO_2. However, gas industry has been started late in China, and there will be no large-scale depleted gas field for a long time. In this way, gas fields should not be used to store CO_2 in the near future but should serve as the strategic energy reserves due to the good sealing property of depleted gas fields [25].

Oil reservoirs are screened on the basis of oil gravity, reservoir temperature and pressure, MMP, and remaining oil saturation, to determine their suitability for CO_2 flooding [17]. And several different types of screening criteria have been proposed at home and abroad for CO_2-EOR and storage [17, 41–44], regarding crude oil properties, reservoir characters, cap formation characters, and economic and environmental issues.

Jilin oilfield, located in northeast of China, is conducting the first large-scale demonstration project on CO_2-EOR and storage. The oil-bearing formations are characterized by good development of sandbody, good connectivity, and well-defined cap rocks [45]. Natural source of CO_2 is mainly from natural gas. And miscible flooding can be achieved in block Hei-59 and Hei-79; well location is indicated in Figure 2.

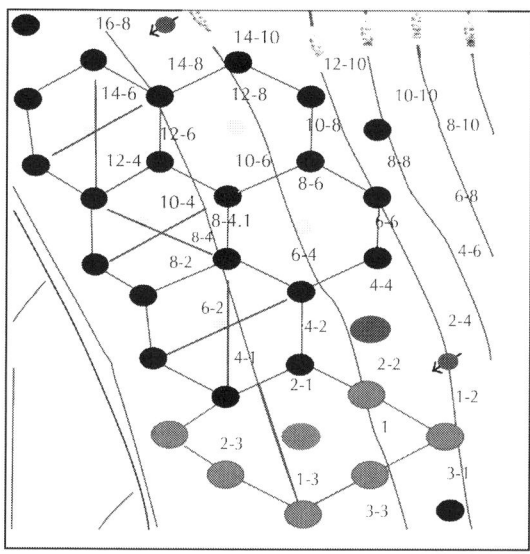

Figure 2: Diagram of well location and surface layout of cross-well seismic lines: yellow dots are CO$_2$ injectors, and the seismic lines are in deep blue color (from Ren et al. [45]).

In 2008, Jilin oilfield built a pilot demonstration area of CO$_2$ flooding and storage in the Daqingzi oilfield. And in 2009, a demonstration area with its annual CO$_2$ storage of 0.2×10^6 t and annual oil displacement of 0.1×10^6 t was established, which indicated realization of commercial application of such technology. Good production response has been observed after about 6 months of CO$_2$ injection since April 2008, as shown in Figure 3. Oil production in the whole pilot area has rapidly increased from 20 t/d to around 100 t/d and has been maintained at 60 t/d in 2011. By the end of May 8, 2011, about 0.167×10^6 t of CO$_2$ was stored without obvious CO$_2$ leakage; and 0.119×10^6 t of oil was produced by CO$_2$-EOR. At the same time, a plant was built in the Jilin oilfield to separate and capture 0.2×10^6 t of CO$_2$ annually [46, 47]. 0.27×10^6 t of CO$_2$ has been safely stored until August 2012 with remarkable economic benefit, with 1:1.37 as the input and output ratio [48]. It is expected that, by 2015, the first production area will be built in China, with an annual CO$_2$ displacement amount reaching 0.5×10^6 t and an annual CO$_2$ storage over 0.7×10^6 t, all of which are equivalent to the total amount of CO$_2$ released from burning of 0.3×10^6 t of coal [46].

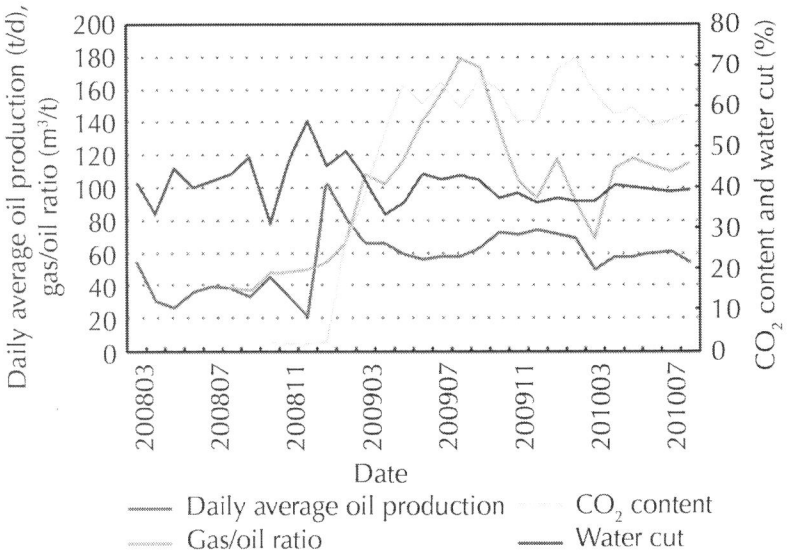

Figure 3: Measured oil production, water cut, CO_2 content and GOR in the CO_2 miscible pilot area of Jinlin oilfield (from Ren et al. [45]).

And the further work will be focused on optimizing EOR performance, verifying of the geo capacity storage in the targeted zones and carrying forward the monitoring programs [45].

Caoshe oilfield is located in Subei Basin and has been selected to implement CO_2-EOR and storage demonstration project. The geological map is shown in Figure 4. Taizhou formation is the main oil-bearing formation in Caoshe oilfield. And during the development periods, the oilfield has developed a complete well pattern of injectors and producers with good well connection, the water cut at the producer has been relatively low, and the reservoir pressure has been well maintained [23, 49].

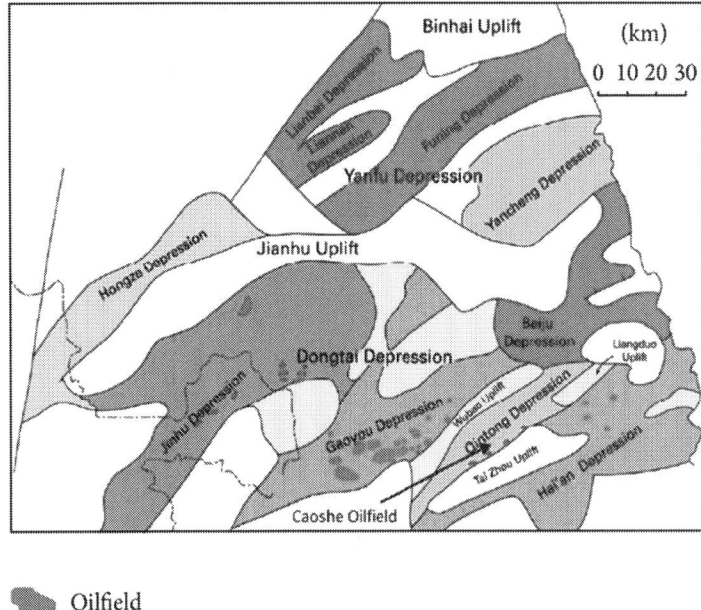

Oilfield

Figure 4: Geotectonic map showing the main depression and uplift regions in the Suibei basin, where the Caoshe oilfield is located (from Zhang [52]).

Taizhou formation is geologically suitable for CSEOR. Taizhou formation has carried out the CO$_2$-EOR pilot test in July 2005, and $5.842 \times 10^7 \, m^3$ CO$_2$ has been injected from July 2005 to December 2009 with increased oil production of 0.03×10^6 t [49, 50]. CO$_2$ can achieve a miscible displacement process and be stored safely in the stratigraphic and structure traps of Taizhou formation reservoir [51]. Besides, Nanjing Chemical plant, a synthetic ammonia plant 120 km away from the Caoshe oilfield, would provide a low-cost CO$_2$ source for the CCS demonstration project. The detailed numerical reservoir model indicates that the maximum CO$_2$ storage capacity at standard condition is estimated to be $0.309 \times 10^9 \, m^3$. Figure 5 shows the simulation result of CO$_2$ miscible flooding. Furthermore, the revenue from incremental oil production is significant, which cannot only offset the cost of the CO$_2$ storage, but also can generate certain economic benefit to Caoshe oilfield [23], while Zhang indicated that the storage cost of CO$_2$-EOR process is \$25.78/t, based on the economic evaluation model established [52].

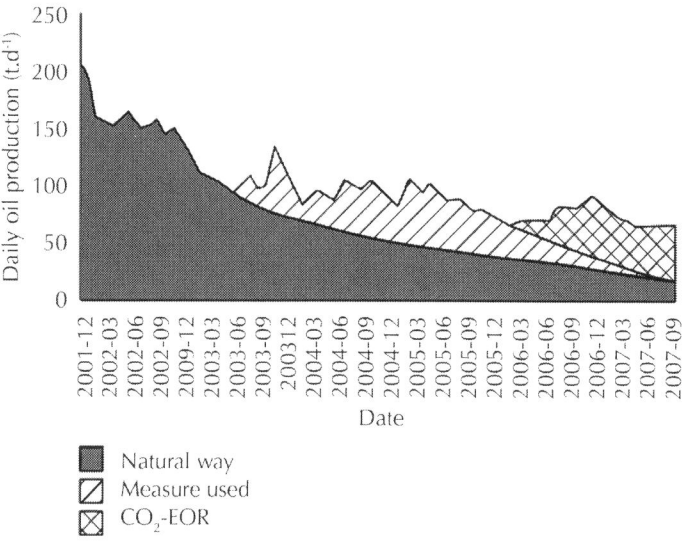

Figure 5: Simulation result of CO_2 miscible flooding of the Taizhou Formation reservoir in the Caoshe oilfield (from Yu et al. [51]).

The Ordos Basin is the second largest sedimentary basin in China, which takes account for 43% of resources of the whole country. In 2011, the oil and gas production exceeded 0.052×10^9 t of oil equivalents [53]. Ordos Basin is able to provide a huge potential capacity for CO_2 storage.

Jingbian field is located in central Ordos Basin in northern Shaanxi slope and has been screened out to conduct the CO_2 sequestration. CO_2 will be captured from the energy and chemical engineering industrial zone in Jingbian City which is 30 km away from the operation site. And it is estimated to inject CO_2 of 0.04×10^6 t/yr and increase oil production of 0.05×10^6 t/yr from CO_2-EOR [53].

Furthermore, various feasibility studies of geological CO_2 sequestration have been implemented for wide areas of Ordos Basin [54–56]. For example, research indicates that, for 261 production layers of Changqing oilfield, total oil production increment and CO_2 sequestration amount can reach about 0.098×10^9 t and about 0.239×10^9 t, respectively [54]. Results from the 50-year injection simulation indicate that a total of 450 Mt of CO_2 can be injected into the targeted reservoir of Majiagou formation (northern Ordos Basin), while 166 Mt of original pore fluids will be displaced by CO_2 [55].

Additionally, other fields around the country also show good results for CO_2 application. Xinjiang oilfield, a vital oilfield in western China, is located in Junggar Basin. Around 0.181×10^9 t additional oil could be produced for the total screened out 275 production units, which could provide about 0.495×10^9 t for CO_2 sequestration capacity [57]. Many mature oil reservoirs in Shengli oilfield (north China) are close to the main CO_2 sources and have good geographical and geological conditions for CO_2 storage. The total EOR potential can be 9.997×10^6 t, and the CO_2 storage capacity can reach 95.539×10^6 t [44]. Zhongyuan oilfield (central China) and Daqing oilfield (northeast China) get obvious recovery increment after CO_2 flooding.

The associated CO_2 from natural gas is another major carbon emission. IPCC estimated that about 50 million tons of reservoir-CO_2 is liberated into the atmosphere every year, from natural gas production [11]. Projecting this to year 2030, and assuming sourness does not increase, the emissions figure could be 150 Mt/yr [58]. And in South China Sea, the geological reserve of CO_2 is huge [59].

DF1-1 gas field is located in the west of the South China Sea, which is associated with a high concentration of CO_2. A demonstrative project of CO_2 sequestration is considered for nearly abandoned southeast block of the lower Group II formation in the DF1-1 gas field, which was reassessed for the safety of CO_2 storage [58]. The separated CO_2 would be injected back into the original gas reservoir, similar to the demonstration projects carried out in K12-B (Netherlands).

The feasibility studies showed that the faults in gas field are characteristic of good sealing property for the targeted block. The injected CO_2 of the southeast block will be effectively trapped in the reservoir because of its good sealing mechanism and poor connectivity with other blocks [60]. Simulation results indicate that CO_2 can be injected steadily at a rate of 0.140×10^6 Sm^3/d over 10 years, and the cumulative CO_2 gas injection can be 0.511×10^9 Sm^3 for the pressure control required. Zhang et al. [60] showed that unit storage of CO_2 is approximately \$20/t at the current economic situation, while there will be no extra finial returns for this demonstration CO_2 sequestration project.

On the other hand, CO_2 injection into oil and gas reservoirs associated with large aquifers takes advantages of lower geological leakage risk from oil and gas traps and large storage capacity from the connected aquifers [61]. Results of cases studies of five oil reservoirs selected from Shengli and Jiangsu oilfields in China demonstrate that CO_2 storage capacity can be greatly increased if the lateral and underlying aquifers are included.

CO$_2$ Sequestration in Saline Aquifer

Deep saline aquifers have proven to be the promising geological media for CO_2 sequestration due to the large storage capacity and wide availability. The injected CO_2 can be sequestrated in deep saline aquifers through a combination of physical and chemical trapping mechanisms, which include stratigraphic or structure trapping, residual trapping, solubility trapping, mineral trapping and hydrodynamic trapping [27, 62–64].

143.505 × 10^9 t CO_2 can be stored in saline aquifers of China [65]. Most of the north China plain; northern, eastern, and southern Sichuan Basin; southeast of Junggar Basin are the priority for CO_2 aquifer storage in the future, like the deep saline aquifers in Songliao Basin (northeast China) can contribute about 8.96 × 10^9 t of CO_2 sequestration capacity [66].

Saline aquifer trap LT13-1, located in the east of DF1-1 gas field, 60 km away from the Dongfang gas terminal, has been selected as the target CO_2 storage site to sequestrate the CO_2 discharged from the DF1-1 gas terminal [67]. The reservoir is relatively good in homogeneousness and high in salinity, indicating a good trap feature. The injected CO_2 will be trapped both in a supercritical state and in dissolved state in formation water. Sandbodies A and C of LT13-1 structure can provide a CO_2 storage capacity of approximately 0.1 × 10^9 t [67], as shown in Figures 6 and 7. Zhang et al. pointed out that the storage cost is about \$33–37/t, slightly higher than abroad due to the high cost of offshore pipeline [68].

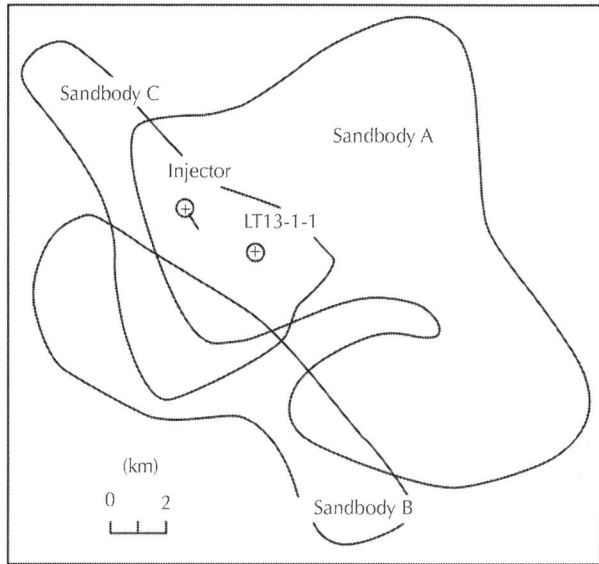

Figure 6: Distribution of sand bodies in the LT13-1 saline aquifer (from Zhang [52]).

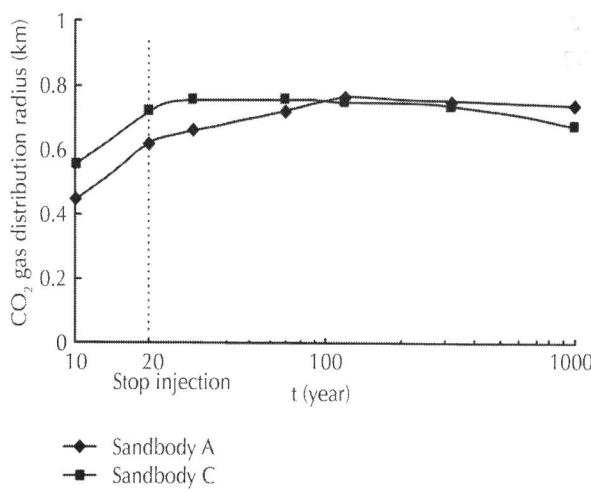

Figure 7: CO$_2$ gas distribution radius in sand bodies A and C during and after injection (from Zhang [52]).

Being one of the most typical sedimentary basins in eastern coastal of China, the Bohai Bay Basin is a potential candidate for CO_2 sequestration. CO_2 storage in deep saline aquifers is considered as a viable option because of the wide-distribution with a high CO_2 storage capacity. The CO_2 storage capacity within the assessing range is 3.9 $\times$ 10^9 t in saline aquifers of Bohai Bay Basin, and storage capacity in Neogene Guantao formation lower than 3500 m is 3.3 $\times$ 10^9 t, accounting for 84.4% of the total potential [70].

Section 3 in the lower part of the Neogene Guantao formation of Beitang Sag, Huanghua depression, near the center of the Bohai Bay Basin, has been chosen as the test site for CO_2 injection [71]. Due to the good cap-rock layers, CO_2 can be stored safely in Section 3 in supercritical state. Based on the model (6) proposed by Tanaka, the CO_2 storage capacity of the Beitang Sag is estimated to be 17.03 Mt.

CO_2 Sequestration in Coal Seam

China has abundant coal bed methane (CBM) resources. CBM reserves buried lower than 2000 m are estimated to be 36.8 Tm^3, accounting for 13% of the world's resources and ranking third in the world [72].

Coal seams provide one of the most attractive sites for CO_2 geological sequestration in China as a result of the huge resources and the high and stable adsorption of CO_2, particularly in combination with ECBM [26, 73, and 74]. Adsorption is the main trapping mechanism for CO_2 storage in coal seams, which accounts for approximately 90% of the total storage. The ECBM potential associated with CO_2 sequestration is estimated to be over 3.751 $\times$ 10^{12} m^3. And the CO_2 sequestration capacity of China coal beds is estimated to be about 142.67 $\times$ 10^9 t [75]. Based on the assessment for coal beds of China in depth between 300 m and 1500 m, 1.632 $\times$ 10^{12} m^3 methane can be increased from CO_2-ECBM, and about 12.078 $\times$ 10^9 t of CO_2 can be stored [26].

The Yaojie coalfield is located in the western margin of Minhe and extends across the Gansu and Qinghai provinces of China. The Haishiwan coalfield is located in the deep part of the Yaojie coalfield. High concentrations of CO_2 (34.1–98.64%) have been observed in the number 2 coal seam of Haishiwan coalfield [69].

And the temperature-pressure conditions in Haishiwan coalfield indicate that supercritical CO_2 may occur in the eastern half of the coalfield. Moreover, the Haishiwan coalfield is an ideal storage area because of the good sealing features and the presence of large volumes of juvenile CO_2 that have been naturally sequestered over 15 million years. The pure CO_2 storage capacity of the Haishiwan coal seam is 44.7 m³/t at 7.5 MPa and 313.15 K [69], as shown in Figure 8.

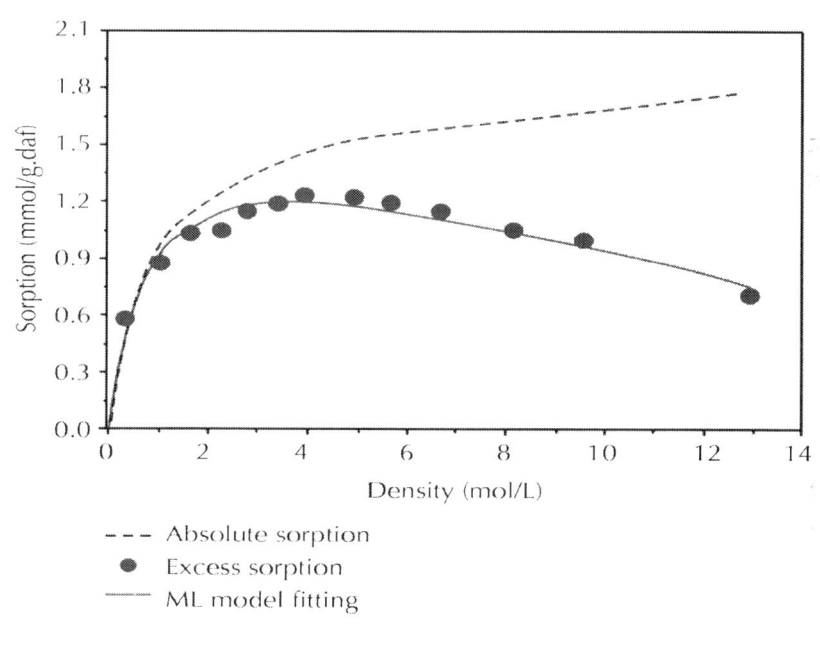

(a)

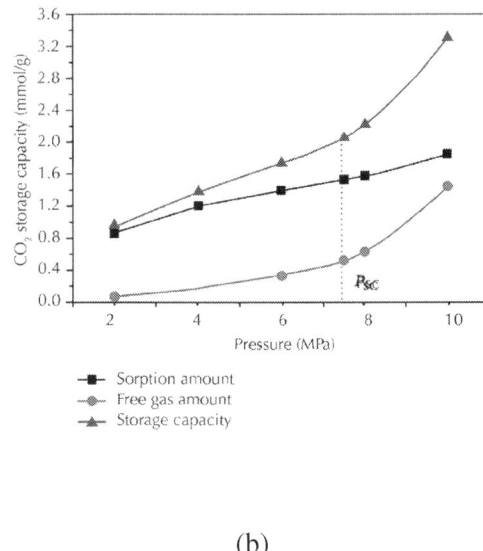

(b)

Figure 8: High-pressure CO_2 adsorption on the dry Haishiwan coals at 40°C with respect to density (a); CO_2 excess sorption isotherms and free CO_2 content versus pressure (b); P_{sc} is the critical pressure of CO_2 (from Li et al. [69]).

Other Ways of Sequestration

Plantation forests are the most effective and ecofriendly way of absorbing CO_2 and increasing carbon sinks in terrestrial ecosystems, mitigating global warming and promoting ecological restoration. China's forestation rate is the highest in the world, contributing significantly to the nation's carbon sequestration [76]. Cost of carbon mitigation through plantation is relatively low, generally under $10/t, compared with $25–120/t for cost limitation of energy industry [77].

China currently has one of the world's most ambitious reforestation and afforestation programs, known as grain for green, which has been in place since 1999. It gives grain payouts to farmers who convert fields to forests. It is operating in many different regions across China. Although not one of its goals, carbon sequestration is a cobenefit of the program [78].

From 1950 to the present, plantations in China sequestered 1.686 PgC by net uptake into biomass and emission of soil organic carbon.

Huang et al. [76] projected that China's forestation activities will continue to net sequester carbon to a level of 3.169 PgC by 2050.

On the other hand, China's rice paddies, accounting for 19% of the world's total, play an important role in soil carbon sequestration. The simulations demonstrated that all the recommended management practices could result in an increase in carbon sequestration potential, varying greatly from 29.2 to 847.7 TgC by 2050 [79].

Additionally, CH_4–CO_2 reforming can effectively convert CO_2 and CH_4 into synthesis gas. Interests regarding the CO_2 reforming of CH_4 have been rising due to the feasible approach for resource utilization and greenhouse gas emission reduction and the generated raw materials needed by many manufacturing process. Many efforts have been carried to devote and investigate various types of catalysts to promote the conversion process [80–84].

Overall, Table 1 summarizes the main comparative information of the above CO_2 sequestration projects regarding different storage ways.

Table 1: Comparison of different CO_2 sequestration projects

Storage media	Total CO2 storage capacity	Project	CO2 storage capacity	EOR potential	Cost of storage
Oilfield	4.6 × 109 t (>800 m)	Jilin	0.7 × 106 t	0.5 × 106 t	1 : 1.37 (input : output)
		Caoshe	0.309 × 109 m3 (by 2009)	0.03 × 106 t (by 2009)	$25.78/t
		Jingbian	0.04 × 106 t/yr	0.05 × 106 t/yr	
		Changqing	t0.098 × 109t	0.239 × 109 t	
		Shengli	t95.539 × 106 t	9.997 × 106 t	
		Xinjiang	0.495 × 109 t		

Gas field	30.483 × 109 t(900–3500 m)	DF1-1	0.511 × 109 Sm3		$20/t
Saline aquifer	143.505 × 109 t	LT13-1	0.1 × 109 t		$33–37/t
		Bohai Bay	3.9 × 109 t		
		Songliao	8.96 × 109 t		
Coal seam	142.67 × 109 t	Haishiwan	44.7/t		
Plantation	3.169 PgC (by 2050)				<$10/t
Soil carbon sequestration	29.2–847.7 TgC (by 2050)				

CHALLENGE FOR FUTURE

CCS is somehow a quite new technology in China. Even though a lot of assessments and potential analysis have been carried out across China, the real commercial implementations are limited. Various factors are supposed to be taken into consideration to promote CO_2 sequestration and to mitigate the deteriorating environment in China.

International engagement is critical in developing and enlarging CO_2 sequestration. China has already cooperated with other countries to start up a number of projects regarding CCS in many fields. However, more combined efforts are needed to move forward.

Technology is the priority determinant in CCS operation, including the technique from capture, transportation, assessment, and storage. The main oilfields in China are manifested in complex formation structure with strong heterogeneity, low or ultralow permeability, low porosity, and poor oil property [1]. CO_2-EOR techniques would be challenged by high miscible pressure, severe gas channeling, heavy solid deposition, and development of complex reservoir [85].

On the other hand, effective policies are suggested to encourage and boost the CCS industry in China. Alternative ways should be developed to capture CO_2 and reduce CO_2 emission for different emitters.

Carbon emission trading system is forming in China. Market mechanism is important to reduce carbon emissions for China's low-carbon future [86].

CONCLUSIONS

The demand for clean energy and low-carbon technologies is enormous in China, where the rapid growth and heavy reliance on coal provide a mass of opportunities for application of new techniques. A great amount of CO$_2$ can be sequestered by geological media, forestation, soil, and reforming. As a result, CCS is the most attractive way for reducing CO$_2$ emission in China.

CO$_2$ sequestration in depleted oil and gas reservoirs, saline aquifers, and coal beds is promising in China. A great number of projects have been implemented to testify the feasibility of CCS, examine the potential for commercial-scale CCS, and assess the storage capacity and possibility of CSEOR in large parts of China like Jilin oilfield, the first large-scale demonstration project on CSEOR.

Forestation, soil, and CO$_2$ reform could provide alternative ways for CO$_2$ sequestration. Combination of variety of methods can deeply promote the emission-reducing work.

There is a gap in carbon sequestration between China and other countries. Besides, most of the CO$_2$ storage projects in China are still in the evaluation and assessment stage. Further efforts are needed to move forward, involving international cooperation, advanced technology, positive policy, and society mechanism.

ACKNOWLEDGMENTS

The authors gratefully acknowledge the contributions of reviewers for critical reading of the paper, constructive comments, and suggestions. This work was supported by National Science Foundation of China (no. 51274173) and National Major Project of China (no. 2011ZX05016-006).

REFERENCES

1. Z. Xuan and S. He, "Potential and early opportunity-analysis on CO_2 geo-sequestration in China," in Proceedings of the 72nd European Association of Geoscientists and Engineers Conference and Exhibition (SPE EUROPEC '10), pp. 842–848, June 2010.

2. S. Miao, "The World's Largest Carbon Emission Reduction Project start in China," Chinese Enterprise News, vol. 1, 2005.

3. S. J. Friedmann, Carbon Capture and Sequestration Technologies: Status and Future Deployment, UCRL-BOOK-235276, 2007.

4. H. Jiang, P. Shen, J. Luo et al., "The present situation and prospect of carbon dioxide storage technology,"Energy and Environment, vol. 32, no. 6, pp. 28–32, 2010 (Chinese).

5. X. Li and Z. Fang, "Status quo of connection technologies of CO_2 geological storage in China," Rock and Soil Mechanics, vol. 28, no. 10, pp. 2229–2239, 2007 (Chinese).

6. A. Saeedi and R. Rezaee, "Effect of residual natural gas saturation on multiphase flow behaviour during CO_2 geo-sequestration in depleted natural gas reservoirs," Journal of Petroleum Science and Engineering, vol. 82-83, pp. 17–26, 2012.

7. M. Loizzo, B. Lecampion, T. Brard, A. Harichandran, and L. Jammes, "Reusing O&G-depleted reservoir for CO_2 storage: pros and cons," in SPE International Petroleum Conference and Exhibition, SPE 124317, 2010.

8. S. R. Ren and L. Zhang, "Geological storage of CO_2: overseas demonstration projects and its implications to China," Journal of China University of Petroleum, vol. 34, no. 1, pp. 93–98, 2010 (Chinese).

9. M. Curtis and S. M. Benson, CO_2 Injection for Enhanced Gas Production and Carbon Sequestration, vol. 74367, SPE, 2002.

10. R. T. Dahowski, C. L. Davidson, X. C. Li, and N. Wei, "A $70/t CO_2 greenhouse gas mitigation backstop for China's industrial and electric power sectors: insights from a comprehensive CCS cost curve,"International Journal of Greenhouse Gas Control, vol. 11, pp. 73–85, 2012.

11. J. Zhang, "CCUS technology and projects of sinopec," in Proceedings of the 4th Annual Global Carbon Capture Utilization

& Storage Summit, pp. 144–163, Beijing, China, 2013.

12.   J. L. Luo, R. Gao, H. Wen-Hui, D. Huo, and Y. Wang, "Carbon dioxide emission reduction and utilization,"Resources & Industries, vol. 13, no. 1, pp. 132–137, 2011.

13.   Intergovernmental Panel on Climate Change (IPCC), IPCC Special Report on Carbon Dioxide Capture and Storage, edited by B. Metz, O. Davidson, Cambridge University Press for the Intergovernmental Panel on Climate Change, Cambridge, UK, 2005.

14.   M. X. Fang, CO$_2$ capture by membrane and adsorption technology, Institute for Thermal Power Engineering of Zhejiang University, 2013.

15.   S. S. Xu, "CO$_2$ capture RD&D projects in China Huaneng Group," in Proceedings of the 4th Annual Global Carbon Capture Utilization & Storage Summit (CCS '13), pp. 102–125, Beijing, China, 2013.

16.   X. M. Song and S. Y. Yang, "Current situation of CCS technology at home and abroad and the positive strategy that China should adopt toward it," Reservoir Evaluation and Development, vol. 1, no. 1-2, pp. 25–30, 2011 (Chinese).

17.   S. Pingping, L. Xinwei, and L. Qiujie, "Methodology for estimation of CO$_2$ storage capacity in reservoirs,"Petroleum Exploration and Development, vol. 36, no. 2, pp. 216–220, 2009.

18.   Y. Zhang, L. Zhang, B. Niu, and S. Ren, "Integrated assessment of CO$_2$-enhanced oil recovery and storage capacity," in Proceedings of the Canadian Unconventional Resources and International Petroleum Conference, October 2010

19.   L. Sun and W. Y. Chen, "Assessment of CO$_2$ geo-storage potential in onshore oil reservoirs, China," China Population, Resources and Environment, vol. 22, no. 6, pp. 76–81, 2012 (Chinese).

20.   S. Tanaka, H. Koide, and A. Sasagawa, "Possibility of underground CO$_2$ sequestration in Japan," Energy Conversion and Management, vol. 36, no. 6–9, pp. 527–530, 1995.

21.   T. Takahashi, T. Ohsumi, and K. Nakayama, "Estimation of CO$_2$ aquifer storage potential in Japan,"Energy Procedia, vol. 1, pp. 2631–2638, 2009.

22. J. Shaw and S. Bachu, "Screening, evaluation, and ranking of oil reservoirs suitable for CO_2-flood EOR and carbon dioxide sequestration," Journal of Canadian Petroleum Technology, vol. 41, no. 9, pp. 51–61, 2002.

23. S. Bachu, J. C. Shaw, and R. M. Pearson, "Estimation of oil recovery and CO_2 storage capacity in CO_2 EOR incorporating the effect of underlying aquifers," in Proceedings of the SPE/DOE Symposium on Improved Oil Recovery, SPE 89340, Tulsa, Okla, USA, April 2004.

24. R. C. Burruss, S. T. Brennan, P. A. Freeman, et al., "Development of a probabilistic assessment methodology for evaluation of carbon dioxide storage," U.S. Geological Survey, Open-File Report 2009-1038, 2009.

25. Y. F. Liu, X. C. Li, Z. Fang, and B. Bai, "Preliminary estimation of CO_2 storage capacity in gas fields in China," Rock and Soil Mechanics, vol. 27, no. 12, pp. 2277–2281, 2006 (Chinese).

26. Y. Liu, X. Li, and B. Bai, "Preliminary estimation of CO_2 storage capacity of coalbeds in China," Chinese Journal of Rock Mechanics and Engineering, vol. 24, no. 16, pp. 2947–2952, 2005.

27. T. Vangkilde-Pedersena, K. L. Anthonsen, N. Smith, et al., "Assessing European capacity for geological storage of carbon dioxide—the EU GeoCapacity project," Energy Procedia, vol. 1, no. 1, pp. 2663–2670, 2009.

28. D. D. Mamora and J. G. Seo, "Enhanced gas recovery by carbon dioxide sequestration in depleted gas reservoirs," in Proceedings of the SPE Annual Technical Conference and Exhibition, pp. 151–159, San Antonio, Tex, USA, October 2002.

29. S. Whittaker, B. Rostron, C. Hawkes, et al., "A decade of CO_2 injection into depleting oil fields: monitoring and research activities of the IEA GHG Weyburn-Midale CO_2 Monitoring and Storage Project," Energy Procedia, vol. 4, pp. 6069–6076, 2011.

30. C. Preston, M. Monea, W. Jazrawi et al., "IEA GHG Weyburn CO_2 monitoring and storage project," Fuel Processing Technology, vol. 86, no. 14-15, pp. 1547–1568, 2005.

31. L. G. H. Van Der Meer, E. Kreft, C. Geel, and J. Hartman, "K12-B A test site for CO_2 storage and enhanced gas recovery," in Proceedings of the 67th European Association of Geoscientists

and Engineers, EAGE Conference and Exhibition, incorporating (SPE EUROPEC '05), June 2005.

32. R. J. Arts, V. P. Vandeweijer, C. Hofstee et al., "The feasibility of CO$_2$ storage in the depleted P18-4 gas field offshore the Netherlands (the ROAD project)," International Journal of Greenhouse Gas Control, vol. 11S, pp. S10–S20, 2012.

33. A. Baklid, R. Korbol, and G. Owren, Sleipner Vest CO$_2$ Disposal, CO$_2$ Injection into a Shallow Underground Aquifer, vol. 36600, SPE, 1996.

34. O. Eiken, P. Ringrose, and C. Hermanrud, "Lessons learned from 14 years of CCS operations: sleipner, in Salah and Snøhvit," Energy Procedia, vol. 4, pp. 5541–5548, 2011.

35. K. Michael, A. Golab, V. Shulakova et al., "Geological storage of CO$_2$ in saline aquifers—a review of the experience from existing storage operations," International Journal of Greenhouse Gas Control, vol. 4, no. 4, pp. 659–667, 2010.

36. S. H. Stevens, D. Spector, and P. Riemer, "Enhanced coalbed methane recovery using CO$_2$ injection: Worldwide resource and CO$_2$ sequestration potential," in Proceedings of the 6th International Oil & Gas Conference and Exhibition in China (IOGCEC '98), pp. 489–501, November 1998.

37. G. Cairns, "Enhanced coal bed Methane recovery and CO$_2$ sequestration in an unmineable coal seam," CONSOL Energy, 2001.

38. E. P. Robertson, "Enhanced coal bed methane recovery and CO$_2$ sequestration in the Powder River Basin," DOE. DE-FC26-05NT42587, 2010.

39. K. V. Veld, C. F. Mason, and A. Leach, "The Economics of CO$_2$ sequestration through Enhanced Oil recovery," Energy Procedia, vol. 37, pp. 6909–6919, 2013.

40. J. Litynski, T. Rodosta, L. Myer, R. Kane, and G. A. Washington, "What is next in geologic CO$_2$ storage research?" in Proceedings of the Carbon Management Technology Conference (CMTC '12), CMTC-151471-MS, Orlando, Fla, USA, 2012.

41. J. J. Taber, F. D. Martin, and R. S. Seright, "EOR screening criteria reyisited—part 1: introduction to screening criteria and enhanced recovery field projects," SPE Reservoir Engineering, vol. 12, no. 3, pp. 189–198, 1997.

42. S. P. Zeng, X. W. Yang, et al., "Fuzzy hierarchy analysis-based selection of oil reservoirs for gas storage and gas injection," Henan Petroleum, vol. 19, no. 4, pp. 40–46, 2005 (Chinese).

43. H. Y. Lei, C. L. Gong, and B. C. Guan, "New screening method for reservoir by CO_2 injection miscible flooding," Journal of China University of Petroleum, vol. 32, no. 1, pp. 72–76, 2008 (Chinese).

44. Z. Liang, W. Shu, Z. Li, R. Shaoran, and G. Qing, "Assessment of CO_2 EOR and its geo-storage potential in mature oil reservoirs, Shengli Oilfield, China," Petroleum Exploration and Development, vol. 36, no. 6, pp. 737–742, 2009 (Chinese).

45. S. Ren, B. Niu, B. Ren et al., "Monitoring on CO_2 EOR and storage in a CCS demonstration project of JiMn Oilfield China," in Proceedings of the SPE Annual Technical Conference and Exhibition 2011 (ATCE '11), pp. 498–505, November 2011.

46. Z. G. Hao, H. C. Fei, and L. Liu, "Integrated techniques of underground CO_2 storage and flooding put into commercial application in the Jilin oilfield," Acta Geologica Sinica, vol. 86, no. 1, p. 285, 2012.

47. "Petrochina's CO_2—EOR Research and Demonstration Project in the Jilin Oil Field (n.d.)," 2014,http://www.ccuschina.org.cn/English/News.aspx.

48. Major R&D Project of "Ji lin Oilfield CCS Technology Research" Passed the Acceptance Inspection, Jilin Xinhuanet, 2012, http://www.jl.xinhuanet.com/2010zhuanti/kanghong/2012   08/23/content_25544196.htm, (Chinese).

49. Z. H. Chen, K. Yu, W. Liu, L. L. Jiao, and M. Yue, "CO_2 miscible flooding and production characteristics and preliminary discussion about the development effects—taking Caoshe oilfield in Northern Jiangsu area as an example," Reservoir Evaluation and Development, vol. 1, pp. 37–41, 2011 (Chinese).

50. F. D. Zhang and Z. L. Wang, "The field experiment and result analysis of CO_2 miscible displacement in Caoshe Oilfield of the North Jiangsu Basin," Petroleum Geology & Experiment, vol. 32, no. 3, pp. 296–300, 2010 (Chinese).

51. K. Yu, W. Liu, and Z. H. Chen, "Study of CO_2 miscible flooding technique in the caoshe oil field, the Qingtong Sag, the Northern

Jiangsu Basin," Petroleum Geology & Experiment, vol. 30, no. 2, pp. 212–216, 2008 (Chinese).

52. Y. Zhang, Research of CO$_2$-EOR and Geological Storage in Jiangsu Caoshe Oilfield, China University of Petroleum, 2010, (Chinese).

53. J. F. Ma, X. Z. Wang, R. Gao, et al., "Monitoring the safety of CO$_2$ sequestration in Jingbian field, China,"Energy Procedia, vol. 37, pp. 3469–3478, 2013.

54. X. W. Liao, C. N. Gao, P. Wu, K. Su, and Y. Shangguan, "Assessment of CO$_2$ EOR and its geo-storage potential in mature oil reservoirs, Changqing Oil Field, China," in Proceedings of the Carbon Management Technology Conference, pp. 62–67, Orlando, Fla, USA, February 2012.

55. Z. S. Jiao, R. C. Surdam, L. Zhoub, P. H. Staufferc, and T. Luob, "A Feasibility Study of Geological CO$_2$ sequestration in the Ordos Basin, China," Energy Procedia, vol. 4, pp. 5982–5989, 2011.

56. X. Ran, Y. Zhao, and X. Liao, "An assessment of a CO$_2$ flood for EOR and sequestration benefits in the Ordos Basin, Northwest China," in Proceedings of the Carbon Management Technology Conference (CMTC '12), 150272, pp. 90–94, February 2012.

57. H. Wang, W. X. Liao, et al., "Potential evaluation of CO$_2$ flooding enhanced oil recovery and geological sequestration in Xinjiang Oilfield," Journal of Shaanxi University of Science & Technology, vol. 31, no. 2, pp. 74–79, 2013 (Chinese).

58. D. Huo and Y. Jalali, "An analysis of CO$_2$ storage prospects in deep saline reservoirs," in Proceedings of the SPE Production and Operations Conference and Exhibition, SPE 136077, Tunis, Tunisia, June 2010.

59. W. Xiang, W. Zhou, J. Zhang et al., "The potential of CO$_2$-EOR in China offshore oilfield," in Proceedings of the SPE Asia Pacific Oil and Gas Conference and Exhibition, pp. 504–507, October 2008.

60. L. Zhang, B. L. Niu, S. Ren et al., "Assessment of CO$_2$ storage in DFI-1 South China sea gas field for CCS demonstration," Journal of Canadian Petroleum Technology, vol. 49, no. 8, pp. 9–14, 2010.

61. L. Zhang, S. Ren, B. Ren, W. Zhang, and Q. Guo, "Assessment of CO$_2$ storage capacity in oil reservoirs associated with large lateral/

underlying aquifers: case studies from China," International Journal of Greenhouse Gas Control, vol. 5, no. 4, pp. 1016–1021, 2011.

62. K. Pruess and T. F. Xu, "Numerical modeling of aquifer disposal of CO_2," in Proceedings of the SPE/EPA/DOE Exploration and Production Environmental Conference, 2001, SPE 66537.

63. J. Moore, M. Adams, R. Allis, S. Lutz, and S. Rauzi, "Mineralogical and geochemical consequences of the long-term presence of CO_2 in natural reservoirs: an example from the Springerville-St. Johns Field, Arizona, and New Mexico, U.S.A," Chemical Geology, vol. 217, no. 3-4, pp. 365–385, 2005

64. M. Wigand, J. W. Carey, H. Schütt, E. Spangenberg, and J. Erzinger, "Geochemical effects of CO_2 sequestration in sandstones under simulated in situ conditions of deep saline aquifers," Applied Geochemistry, vol. 23, no. 9, pp. 2735–2745, 2008.

65. X. C. Li, Y. F. Liu, B. Bai, and Z. Fang, "Ranking and screening of CO_2 saline aquifer storage zones in China," Chinese Journal of Rock Mechanics and Engineering, vol. 25, no. 5, pp. 963–968, 2006 (Chinese).

66. L. M. Liao and J. Wang, "Study of geological storage of carbon dioxide in the Songliao Basin," Geological Journal of Sichuan, vol. 32, no. 3, pp. 268–271, 2012 (Chinese).

67. L. Zhang, Y. Zhang, H. G. Mi, S. R. Ren, and Y. X. Ma, "CO_2 storage in saline aquifers: design of a demonstration project to dispose CO_2 associated with natural gas fields in South China sea," inProceedings of the Canadian Unconventional Resources and International Petroleum Conference, pp. 95–101, Alberta, Canada, October 2010.

68. L. Zhang, S. Ren, R. Wang, P. Yi, H. Mi, and J. Li, "Feasibility study on associated CO_2 geological storage in a saline aquifer for development of Dongfang 1-1 gas field," Journal of China University of Petroleum, vol. 34, no. 3, pp. 89–93, 2010.

69. W. Li, Y. Cheng, L. Wang, H. Zhou, and H. Wang, "Evaluating the security of geological coalbed sequestration of supercritical CO_2 reservoirs: the Haishiwan coalfield, China as a natural analogue,"International Journal of Greenhouse Gas Control, vol. 13, pp. 102–111, 2013

70. Z. H. Pang, "Study on CO$_2$ sequestration in saline aquifers in the Bohai Bay Basin," in Proceedings of the 4th Annual Global Carbon Capture Utilization & Storage Summit, pp. 769–782, 2013.

71. Z. Pang, Y. Li, F. Yang, and Z. Duan, "Geochemistry of a continental saline aquifer for CO$_2$ sequestration: the Guantao formation in the Bohai Bay Basin, North China," Applied Geochemistry, vol. 27, no. 9, pp. 1821–1828, 2012.

72. "Natural Gas Industry" Editorial Department, "CBM resources in China ranks third in the world,"Natural Gas Industry, vol. 30, no. 5, p. 128, 2010.

73. Z. Fang, X. Li, H. Li, and H. Chen, "Feasibility study of gas mixture enhanced coalbed methane recovery technology," Rock and Soil Mechanics, vol. 31, no. 10, pp. 3223–3229, 2010 (Chinese).

74. S. D. Golding, I. T. Uysal, C. J. Borehama, D. Kirstea, K. A. Baublysb, and J. S. Esterle, "Adsorption and mineral trapping dominate CO$_2$ storage in coal system," Energy Procedia, vol. 4, pp. 3131–3138, 2011. View at Publisher · View at Google Scholar

75. H. Yu, G. Zhou, W. Fan, and J. Ye, "Predicted CO$_2$ enhanced coalbed methane recovery and CO$_2$sequestration in China," International Journal of Coal Geology, vol. 71, no. 2-3, pp. 345–357, 2007.

76. L. Huang, J. Y. Liu, Q. Shao, and X. Xu, "Carbon sequestration by forestation across China: past, present, and future," Renewable and Sustainable Energy Reviews, vol. 16, no. 2, pp. 1291–1299, 2012.

77. W. L. Wang, "Plantation—the effective way for low-carbon life," Fujian Daily, vol. 3, 2013 (Chinese).

78. I. M. Caldwell, V. W. Maclaren, J. M. Chen et al., "An integrated assessment model of carbon sequestration benefits: a case study of Liping county, China," Journal of Environmental Management, vol. 85, no. 3, pp. 757–773, 2007.

79. S. X. Xu, X. Z. Shi, Y. Zhao et al., "Carbon sequestration potential of recommended management practices for paddy soils of China, 1980–2050," Geoderma, vol. 166, no. 1, pp. 206–213, 2011.

80. G. J. Zhang, Y. Dong, M. R. Feng, Y. Zhang, W. Zhao, and H. Cao, "CO$_2$ reforming of CH$_4$ in coke oven gas to syngas over coal

char catalyst," Chemical Engineering Journal, vol. 156, no. 3, pp. 519–523, 2010.

81. G. J. Zhang, Y. N. Du, Y. Xu, and Y. F. Zhang, "Effects of preparation methods on the properties of cobalt/carbon catalyst for methane reforming with carbon dioxide to syngas," Journal of Industrial and Engineering Chemistry, vol. 20, no. 4, pp. 1677–1683, 2014

82. G. J. Zhang, J. W. Qu, and A. T. Su, "Towards understanding the Carbon catalyzed CO_2 reforming of Methane to syngas," Journal of Industrial and Engineering Chemistry, 2014.

83. X. P. Yu, N. Wang, W. Chu, and M. Liu, "Carbon dioxide reforming of methane for syngas production over La-promoted NiMgAl catalysts derived from hydrotalcites," Chemical Engineering Journal, vol. 209, pp. 623–632, 2012.

84. Y. Zhang, G. Zhang, Y. Zhao, X. Li, Y. Sun, and Y. Xu, "Ce-K-promoted Co-Mo/Al_2O_3 catalysts for the water gas shift reaction," International Journal of Hydrogen Energy, vol. 37, no. 8, pp. 6363–6371, 2012

85. X. A. Yue, R. B. Zhao, and F. L. Zhao, "Technological challenges for CO_2 EOR in China," Science paper Online, vol. 2, no. 7, pp. 487–491, 2007 (Chinese).

86. R. Y. Li, "Report on the Development of Low Carbon Economy of China," Released. GMW, 2013,http://politics.gmw.cn/2013-05/27/content_7767951.htm.

Chapter 7

An Experimental and Modeling Study on the Response to Varying Pore Pressure and Reservoir Fluids in the Morrow a Sandstone

Aaron V. Wandler, Thomas L. Davis,
and Paritosh K. Singh

Department of Geophysics, Colorado School of Mines, 1500 Illinois
Street, Golden, CO 80401, USA

ABSTRACT

In mature oil fields undergoing enhanced oil recovery methods, such as CO_2 injection, monitoring the reservoir changes becomes important. To understand how reservoir changes influence compressional wave (P) and shear wave (S) velocities, we conducted laboratory core experiments on five core samples taken from the Morrow A sandstone at Postle Field, Oklahoma. The laboratory experiments measured P- and S-wave velocities as a function of confining pressure, pore pressure, and fluid type (which included CO_2 in the gas and supercritical phase). P-wave velocity shows a response that is sensitive to both pore pressure and fluid saturation. However, S-wave velocity is primarily sensitive to changes in pore pressure. We use the fluid and pore pressure response measured from the core samples to modify velocity well logs through a log facies model correlation. The modified well logs simulate the brine- and CO_2-saturated cases at minimum and maximum reservoir pressure and are inputs for full waveform seismic modeling. Modeling shows how P- and S-waves have a different time-lapse amplitude response with offset. The results from the laboratory experiments and modeling show the advantages of combining P- and S-wave attributes in recognizing the mechanism responsible for time-lapse changes due to CO_2 injection.

INTRODUCTION

Currently, there are 128 enhanced oil recovery (EOR) projects worldwide using CO_2 injection as a tertiary recovery method, with 113 taking place in the United States [1]. Often the criterion for successful seismic monitoring of CO_2 projects is the ability to collect high-resolution seismic data, since seismic methods have shown to be very effective in detecting time-lapse changes [2–6]. The effectiveness of seismic monitoring depends upon the changes in acoustic and elastic impedance caused by injection and production. These changes are then observed in the amplitude and travel time of seismic waves.

Time-lapse seismic methods utilizing compressional waves (P-wave) and shear waves (S-wave) can ideally improve the identification of flooded or pressurized compartments within the reservoir during a CO_2 EOR operation. The use of P-wave data is not a definitive method for pressure detection, but the addition of S-wave data may provide a distinction between a gas-bearing zone and a zone of high pore pressure [7], this is a motivation for using S-waves as a time-lapse tool. An example provided by Xue et al. [8] explains if an observed time-lapse P-wave anomaly is caused by a decrease in velocity without an associated S-wave anomaly, the zone has most likely been flushed with CO_2 without any changes in pore pressure. However, if the S-wave data show a similar anomaly, due to a reduction in velocity, the zone may have undergone a pore pressure buildup in addition to being flushed with CO_2. Identifying the impact of changing stresses within the reservoir is as important as recognizing the effects of saturation.

Saturation effects are typically estimated from fluid substitution modeling (e.g., Gassmann [9] and Kuster and Toksoz [10]). Since fluid substitution methods do not directly account for changing rock properties with respect to varying pore pressure, additional information must be used to account for the effects of pore pressure. This additional information may be obtained from, for example, reservoir flow simulation or experimental core studies. The purpose of this paper is to show how the results from rock physics experiments that include saturation and pore pressure variation can be used to estimate amplitude changes in a reservoir undergoing CO_2 injection.

The contribution of this study applies directly to a major ongoing research project at Colorado School of Mines. The Reservoir Characterization Project (RCP) is a consortium at Colorado School of Mines whose research goal is to monitor CO_2 EOR projects to effectively manage and enhance the recovery process. The RCP has used P-, S-, and converted-wave (PS) time-lapse seismic acquisition to monitor CO_2 floods for EOR. The RCP conducted the first ever time-lapse, multicomponent seismic survey in conjunction with a CO_2 huff-n-puff project at Vacuum Field, New Mexico. Multicomponent time-lapse seismic data was used to monitor the EOR project at Weyburn Field, Saskatchewan, Canada. The current research project is a time-lapse study of an incised valley fill system undergoing a water alternating gas (i.e., CO_2) injection scheme for EOR at Postle Field, Texas County, Oklahoma, located in the Anadarko Basin. The results from the experiment and analysis reflect the use of reservoir core samples and actual reservoir fluids. The aim of this work is to provide insight related directly to this specific type of environment.

Previous Work

Studies on the effects of saturation, pressure, and pore fluid have been documented as early as 1966 [12]. King [12] reported on the change in velocities of five sandstone samples saturated with air, an NaCl solution, and kerosene as a function of hydrostatic confining pressure and internal pore pressure. Furthermore, Domenico [13] showed that S-wave velocity is more sensitive to increasing differential pressure than P-wave velocity in three sandstone cores of low, intermediate, and high porosity.

A series of papers have been published on the results of the P-wave velocity response to the migration of CO_2 in water-saturated sandstones using an experimental setup that simulates cross-well seismic profiling. Xue et al. [8] mapped the velocity changes caused by CO_2 injection in water-saturated Tako and Shirahama sandstones at constant pore pressure. In a similar experiment, Xue and Ohsumi [14] mapped the velocity changes caused by the injection of gas, liquid, and supercritical phase CO_2 into a water-saturated Tako sandstone. They investigated the changes in velocity with increasing pore pressure and concluded that CO_2 injection causes larger velocity changes than the change associated with increasing pore pressure in their core samples. In addition to the P-wave velocity, strain was measured as a function of hydrostatic pressure in a dry sample. Their observations showed how strain normal and parallel to the bedding plane are not equal and more deformation occurred normal to the bedding plane. To improve upon their cross-well seismic profiling experiment and the accuracy of the P-wave velocity changes relative to the CO_2 injection, Xue and Lei [15] applied a difference tomography method to the measured slowness. This technique correlated areas of higher porosity with greater velocity reduction. Velocity reduction when CO_2 was in liquid or supercritical phase was twice as much as when in the gas phase. From the same dataset as Xue and Lei [15], Shi et al. [16] focused on P-wave velocity reductions due to supercritical CO_2 displacement of pore water, where the velocity reductions deviated significantly from the Gassmann-predicted velocities for both patchy and uniform saturations. Lei and Xue [17] expanded on the results of Xue and Lei [15] and Shi et al. [16] and showed that P-wave velocity decreased, and the attenuation coefficient increased, due to gaseous, liquid, and supercritical CO_2 partially replacing pore water. Kim et al. [18] also investigated the use of P-wave velocity and resistivity as a way to monitor saturation of a sandstone during a CO_2 injection process. This was accomplished by injecting supercritical CO_2 into a water-saturated Berea sandstone sample at simulated reservoir conditions.

The effects of saturation, pore pressure, and effective stress, studied on sandstone core samples from the Northwest Shelf of Australia, show that the V_p/V_s ratio is an indicator for the saturation state between oil-saturated and the dry state [19]. Siggins [20] also studied the response of CO_2-saturated sandstones at in-situ reservoir conditions for two synthetic sandstones and one reservoir sandstone/mudstone from the Otway Basin, C seam, Waarre Formation. His findings were both P-wave and S-wave velocities decreased when the samples were saturated with liquid CO_2, compared to dry rock. Also, the synthetic sandstones showed better agreement with Gassman's fluid substitution theory than did the reservoir sandstone. Similarly, Siggins et al. [21] compared synthetic sandstones and Waarre C formation reservoir sandstones when dry and saturated with gaseous and liquid-phase CO_2. Larger changes were observed in both P-wave and S-wave velocity when CO_2 was in the liquid phase.

The effects of saturation and pore pressure have also been studied with respect to the effective pressure law and the effective pressure coefficient [22, 23] and how fractures change the constituency of the rock and the effective pressure coefficient [24]. The work by Xu et al. [24] on effective pressure shows the importance of why assuming the effective pressure coefficient to be unity could lead to significant errors when predicting time-lapse changes.

The RCP has used experimental rock physics to understand how the changes in confining pressure, pore pressure, and fluid saturation affect P-wave and S-wave velocities. This knowledge has then been applied to the interpretation of multicomponent time-lapse surface seismic monitoring. Capello [25] used rock physics as a feasibility study for seismic monitoring of a CO_2 huff-n-puff flood in the Permian San Andres Formation, a dolomitized carbonate deposit at Vacuum Field. Also using samples from the San Andres Formation, Duranti [26] investigated the coexistence of equant pores and fractures and found that a dual porosity rock physics model was more successful in predicting the changes in differential pressure and saturation observed in the experimental data.

Brown [27] integrated experimental rock physics measurements from the carbonate Marly beds of the Mississippian Charles Formation to improve the interpretation of time-lapse seismic data for the monitoring of a CO_2 flood at Weyburn Field. Yamamoto [28] included the rock physics results obtained by Brown [27] to improve the flow modeling of CO_2 by minimizing the difference between calculated and observed acoustic impedance while simultaneously matching the production history.

Lastly, Rojas [29] measured dry rock velocities for the purpose of analyzing the different responses from Gassmann fluid substitution. This study on tight gas sands linked V_p/V_s sensitivity to pore pressure changes, lithology, and fluid content to aid in identifying zones with a higher reservoir potential of the Late Cretaceous Williams Fork Formation at Rulison Field, CO.

Perhaps the earliest and most cited laboratory experiment related to monitoring a CO_2 EOR operation is that by Wang and Nur [30]. They measured ultrasonic P-wave and S-wave velocities on seven sandstones and one unconsolidated sand saturated with n-hexadecane and then flushed with CO_2. Their measurements were done at constant confining pressure over a range of pore pressures and temperatures above and below the critical point of CO_2. Their findings show a significant decrease in P-wave velocity which was attributed to CO_2 saturation. The S-wave velocity exhibited a greater effect to pore pressure than saturation and was shown to be less sensitive to saturation.

Outside of research done on the Tako, Shirahama, Waarre C Formation, Williams Fork Formation sandstones and the carbonates of San Andres and Charles Formation, most of the experimental studies are not done on reservoir samples obtained from well core. This is evident in the statement made by Siggins et al. [21], "literature on laboratory measurements of the effects of CO_2 on the seismic properties of reservoir rocks is relatively sparse."

POSTLE FIELD OVERVIEW

The geologic environment of the producing reservoir at Postle Field is the Upper Morrow sandstone Formation, which is a Pennsylvanian-age siliciclastic incised valley fill system. Typical Morrow sediments consist of shales punctuated by valley fill sand deposits [31]. The Morrow Formation is encased between two high-stand limestones [32]. The upper boundary is a conformable surface at the base of the Atoka Thirteen Fingers Limestone. The lower boundary of the Morrow Formation is an unconformable surface above the Mississippian Chester Limestone. The deposition of Postle Field during Pennsylvanian time was a relative sea-level low-stand shoreline where the Morrow Formation is much sandier [33]. The primary producing unit in the Morrow Formation is the Upper Morrow A, A1, and A2 sands. In Postle Field, the producing formation that is undergoing the CO_2 flood is the Morrow A sandstone.

Postle Field is a large mature oil field undergoing a water alternating gas (WAG) flood for enhanced oil recovery of the Morrow A sandstone. The volume of water to CO_2 injected into the reservoir is referred to as the WAG ratio and is expressed in reservoir volumes at reservoir conditions. The average WAG ratio implemented at Postle Field is 0.35 to 1. The reservoir pressure is maintained above the minimum miscibility pressure of 14.48 MPa by injecting water and CO_2 at 29.65 MPa and by keeping the bottom hole pressure of a producing well higher than 6.89 MPa [34].

SAMPLE SELECTION

We retrieved five core plug samples from three different lithological zones within the Morrow a sandstone interval. The three zones were a low-porosity and low-permeability cemented zone, a higher-porosity and high-permeability zone, and a higher-porosity and lower-permeability zone. Each sample was cut parallel to the bedding plane. The cylindrical dimensions of the samples were roughly 3.81 cm in length and 2.54 cm in diameter. The depths from where the core samples were retrieved, as well as density, porosity, and permeability are listed in Table 1.

Table 1: Depth, density, porosity, and permeability of core samples from three different lithological zones

Sample	Depth (m)	Density (gm/cm³)	Porosity (%)	Permeability (mD)	Permeability (mD)
				Preexperiment	Postexperiment
Cemented	1875	2.394	11.54	10.6	6.67
High permeability no. 1	1878	2.105	19.96	270	192

High permeability no. 2	1879	2.067	22.64	123	105
Lower permeability no. 1	1888	2.179	18.73	11.0	11.0
Lower permeability no. 2	1888	2.104	20.84	68.8	26.8

Mineralogy

We used QEMSCAN (quantitative evaluation of minerals by scanning electron microscopy) to quantitatively evaluate mineral content on each of the three lithological zones. The cemented zone (Figure 1) consists of subrounded quartz grains cemented by ankerite and kaolinite clay. This cemented zone shows a lower amount of grain-to-grain contact which suggests the possibility of early cementation. The lower amount of grain-to-grain contact is likely related to the presence of ankerite which may indicate that the pore water was rich in iron. In addition, there are occurrences of angular overgrowths over subrounded quartz grains. Even though the QEMSCAN measured 14% porosity, most of the porosity for this cemented zone is interstitial between the fine-grained kaolinite clay.

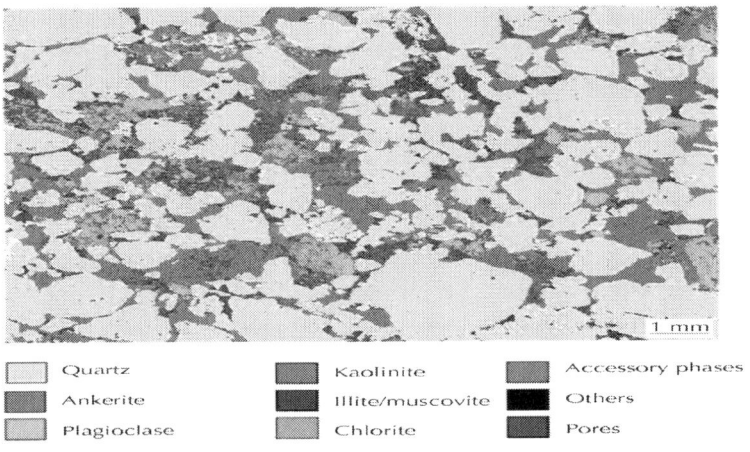

Quartz · Kaolinite · Accessory phases
Ankerite · Illite/muscovite · Others
Plagioclase · Chlorite · Pores

Figure 1: QEMSCAN image from the cemented zone.

The higher-porosity and high-permeability zone (Figure 2) consists mostly of subrounded quartz grains with lesser amounts of kaolinite clay cement, where most of the porosity (25%) occurs between quartz grains. Unlike the other zones, only a trace amount of Ankerite occurs as cement.

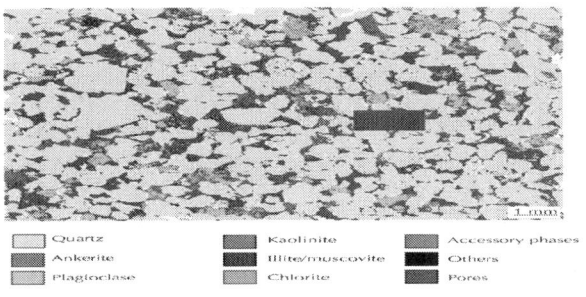

Figure 2: QEMSCAN image from the high-permeability zone. (The solid blue rectangle is a file conversion artifact).

The higher-porosity and lower-permeability zone (Figure 3) consists mostly of subrounded quartz grains with lesser amounts of ankerite and kaolinite clay cement. This zone is more similar to the cemented zone with respect to the presence of ankerite and interstitial porosity. Much like the cemented zone, most of the porosity (20%) is interstitial between the fine-grained kaolinite and chlorite.

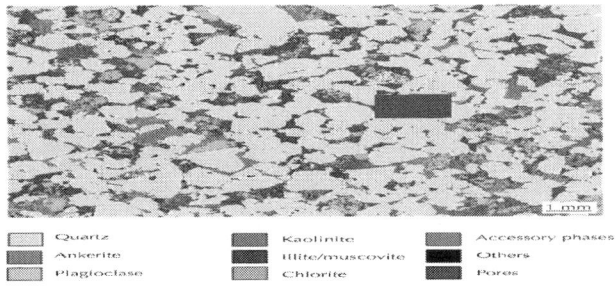

Figure 3: QEMSCAN image from the lower-permeability zone. (The solid blue rectangle is a file conversion artifact).

Table 2: shows the mineral volume percentage related to each of the zones as described above.

Table 2: QEMSCAN mineral volume (%) of the three zones

Zone	Quartz	Ankerite	Plagioclase	Kaolinite	Illite	Chlorite	Accessory Phases	Others
Cemented	66	19	6	7	trace	trace	trace	1
High permeability	84	trace	6	8	1	trace	trace	trace
Lower permeability	75	4	8	8	1	3	trace	trace

METHODOLOGY

Experimental Description

For the laboratory core experiments, we used the direct pulse transmission technique to record compressional and shear-wave velocities at ultrasonic frequencies (≈ 1 MHz). The experimental setup included a pressure vessel to encase and apply an isostatic confining pressure to the core sample; heat tapes wrapped around the pressure vessel to control the temperature; independent pumps for regulating confining pressure and pore pressure; a pulse generator; a digital oscilloscope for recording wave forms. A transducer was placed at each flat end of the core plug sample and was capable of recording a compressional wave and two shear waves in orthogonal directions. The two shear waves were oriented along the fast- and slow-velocity directions relative to the core sample at bench-top conditions. Each transducer casing had a fluid line for injecting the different fluids through the sample. In addition to recording the different wave modes, average axial strain was measured along the major axis of the sample. Uncertainties from the experimental setup correspond to a velocity error of 0.3% to 0.4%. However, the actual error is higher due to hand-picking first arrivals.

Testing Sequence

We started the testing sequence with dry rock measurements where we incrementally increased the confining pressure from 3.45 MPa to 41.37 MPa at the estimated reservoir temperature of 63.9°C.

Following the dry rock measurements, we vacuum saturated the core sample with brine then recorded measurements at different pore pressures. The fluids flushed through the samples following brine were live oil, live oil with a 0.334 mol fraction of CO_2, and pure CO_2. The pore pressure range for the brine-saturated samples was 29.65 MPa to 3.45 MPa, to simulate the stress path from an injection well to a production well. For the other fluids, live oil, live oil with a 0.334 mol fraction of CO_2, and pure CO_2, the pore pressure range was 29.65 MPa to 6.89 MPa. For these fluids, we did not record data below 6.89 MPa since these pressures would be below the bubble point. For the fluid-saturated and fluid-flushed cases, a constant confining pressure of 44.82 MPa is maintained to stimulate the stress of the overburden, and all the measurements were recorded at 63.9°C.

Fluid Properties

The fluids used to saturate and flush through the core samples were reproduced to simulate the actual reservoir fluids at Postle Field. These fluids consist of brine with an NaCl concentration of 142,000 ppm, live oil with a gas oil ratio (GOR) of 103 L/L, live oil with a 0.334 mole fraction of CO_2, and pure CO_2. The live oil has an API gravity of 38.6 and a gas gravity of 0.786. This information was obtained from a reservoir fluid study provided by the current field operator. Using the Fluid Acoustics for Geophysics (FLAG) calculator [11], the live oil bubble point is estimated to be 13.33 MPa at 63.9°C. The mixture consisting of live oil and a mole fraction of 0.334 CO_2 has an estimated bubble point of 17.14 MPa at 63.9°C. The GOR and volume of gas for the live oil with a 0.334 mole fraction of CO_2 was calculated assuming a gas gravity of 1.53 and a molecular weight of 44.01 gm/mole for CO_2. This approach was used by Wang et al. [35] based on experimental measurements of Postle Field oil and mixtures of approximately 10 mole% and 50 mole% CO_2. Figures 4 and 5 show the properties of the fluids over the pore pressure range used in the experiment calculated with the FLAG calculator [11]. The phase of CO_2 for these pressures is supercritical fluid with the exception of the lowest pressure, 6.89 MPa, where CO_2 is in the gas phase.

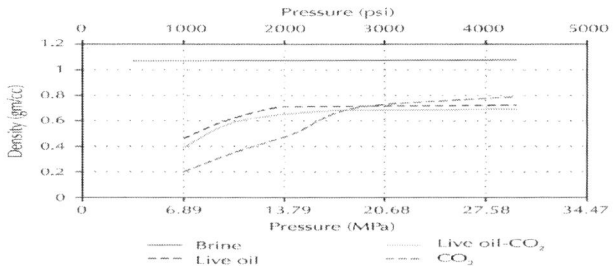

Figure 4: Fluid density as a function of pressure, modeled using the FLAG calculator [11].

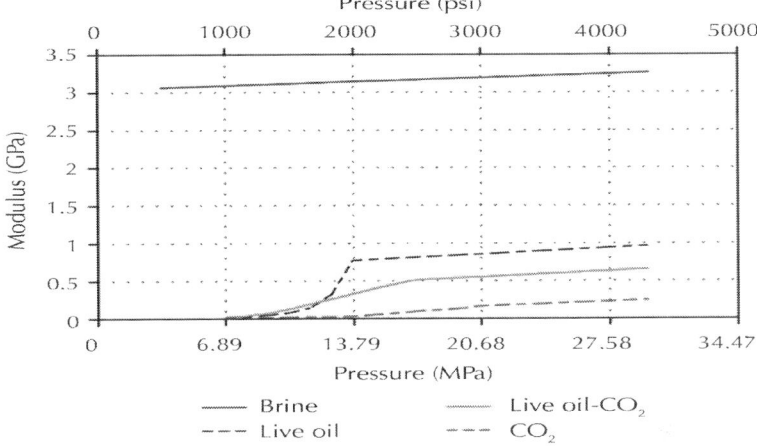

Figure 5: Bulk modulus of fluids as a function of pressure, modeled using the FLAG calculator [11].

RESULTS AND DISCUSSION

First, we measured compressional wave (P-wave) and shear wave (S-wave) velocities in air-saturated samples (dry rock) while loading and unloading an isostatic confining pressure. Following the dry rock testing, we measured P- and S-wave velocities, as a function of pore pressure, on brine-saturated core samples which were then flooded with live oil, live oil with a 0.334 mol fraction of CO_2, and pure CO_2, with a constant confining pressure of 44.82 MPa.

Core-Measured Velocities

The measured velocities, P- and S-wave, are shown in Figures 6–10 for the dry rock and fluid-saturated core samples. The S-wave velocity shown in these figures is the average of the two orthogonally measured S-waves. Differential pressure, P_d, is defined as confining pressure, P_c, minus pore pressure, P_p: (1);

$$P_d = P_c - P_p \qquad\qquad (1)$$

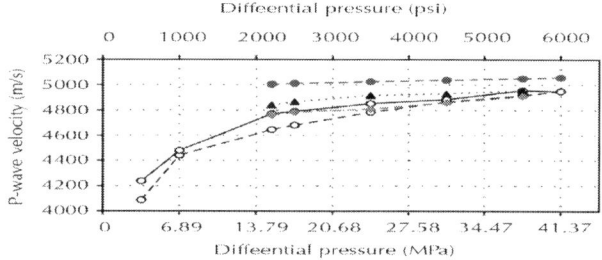

Figure 6: P- and S-wave velocities for the cemented sample versus differential pressure for dry rock, brine saturated, and when flushed with live oil and CO_2.

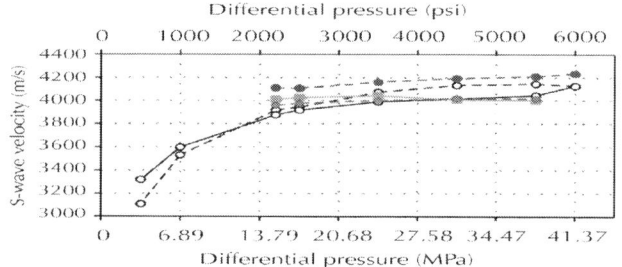

Figure 7: P- and S-wave velocities for the high-permeability number 1 sample versus differential pressure for dry rock, brine saturated, and when flushed with the oil-CO_2mixture, and CO_2.

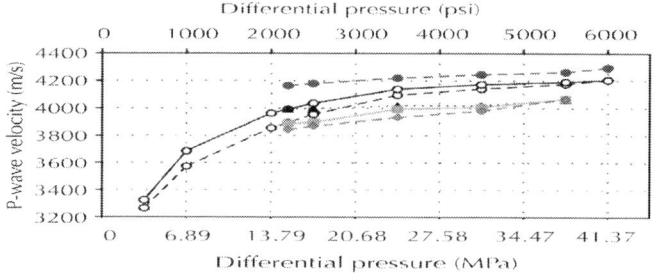

Figure 8: P- and S-wave velocities for the high-permeability number 2 sample versus differential pressure for dry rock, brine saturated, and when flushed with live oil, the oil-CO_2 mixture, and CO_2.

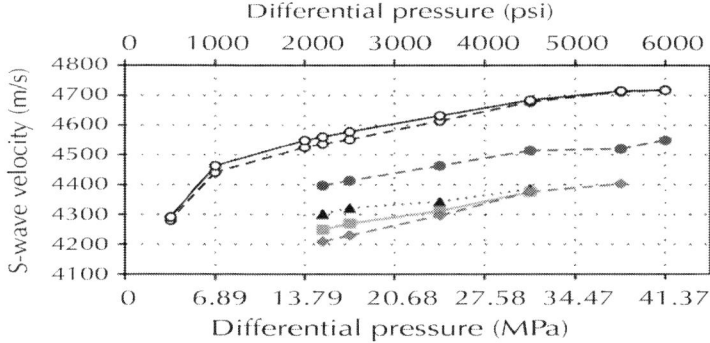

Figure 9: P- and S-wave velocities for the low-permeability number 1 sample versus differential pressure for dry rock, brine saturated, and when flushed with live oil, the oil-CO_2 mixture, and CO_2. (Note dry rock velocities were recorded at 20°C, fluid-saturated velocities were recorded at 63.9°C).

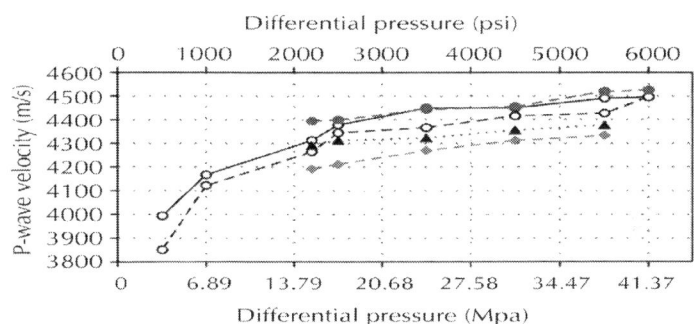

Figure 10: P- and S-wave velocities for the low-permeability number 2 sample versus differential pressure for dry rock, brine saturated, and when flushed with live oil and CO_2.

In general, the high-permeability samples show a greater sensitivity to both confining pressure and pore pressure. Both P-wave and S-wave velocities decrease with decreasing differential pressure (i.e., increasing pore pressure). The P-wave velocities show a larger decrease in velocities due to pressure when flooded with CO_2. Whereas pore fluid has little bearing on S-wave velocities. Tables 3 and 4 show the percentage change in velocity for dry rock and fluid-saturated core samples over the differential pressure range of 15.17 MPa to 37.92 MPa.

Table 3: Percentage change of P-wave velocity with increasing differential pressure from 15.17 MPa to 37.92 MPa

Sample	Dry Rock Loading (%)	Dry Rock Unloading (%)	Brine Sat. (%)	Oil Flushed (%)	Oil-CO$_2$ Flushed (%)	CO$_2$ Flushed (%)
Cemented	6.5	−3.6	0.94	2.4	—	3.2
High permeability no. 1	5.4	−6.1	2.5	—	−0.24	1.4
High permeability no. 2	8.0	−5.1	2.4	1.8	4.6	5.9
Lower permeability no. 1	4.0	−3.3	2.8	1.9	3.0	4.6
Lower permeability no. 2	5.4	−4.1	2.9	2.0	—	3.4

Table 4: Percentage change of S-wave velocity with increasing differential pressure from 15.17 MPa to 37.92 MPa

Sample	Dry Rock Loading (%)	Dry Rock Unloading (%)	Brine Sat. (%)	Oil Flushed (%)	Oil-CO$_2$ Flushed (%)	CO$_2$ Flushed (%)
Cemented	6.5	−4.3	2.3	2.3	—	2.9
High permeability no. 1	10.2	−8.3	7.7	—	10.5	7.7
High permeability no. 2	8.3	−6.4	8.7	9.7	10.3	9.1
Lower permeability no. 1	6.4	−5.2	5.0	5.9	5.3	5.7
Lower permeability no. 2	6.9	−5.5	3.4	3.9	—	4.4

In terms of different lithological characteristics of the core samples with respect to velocity, the higher the permeability of the sample, the lower the P-wave and S-wave velocities. This relationship cannot be extended to porosity since the porosities of the lower-permeability samples were very to the high-permeability samples.

The P-wave velocities show a response to both pressure and fluid type. The P-wave velocities systematically decrease, where brine saturated is the fastest and CO_2 flushed is usually the slowest. The S-wave velocities' response is mainly dominated by pressure and generally the fluid effect is negligible.

Some specific observations with respect to each of the core samples are as follows.

Cemented Sample

The S-wave velocity for the brine-saturated sample has a slightly higher velocity than when CO_2 flushed (Figure 6(b)). More commonly the opposite behavior is observed.

High Permeability Number 1

The dry rock P-wave velocities (Figure 7(a)) show an unusual pattern where the velocity due to loading is higher than the velocity while unloading. This loading and unloading cycle may have caused permanent damage to the core leading to this response. Additionally, the P-wave velocities of the mixture consisting of live oil with a mole fraction of $0.334\,CO_2$ and pure CO_2 become nearly equal at a differential pressures of 31.03 MPa and 37.92 MPa (equivalent to pore pressures of 13.79 MPa and 6.89 MPa). The estimated bubble point for the oil-CO_2 mixture is 17.14 MPa, suggesting the dominant phase of the mixture at these pressures could be gas.

In regards to the S-wave velocities (Figure 7(b)), the dry rock velocities do not show the same pattern as the P-wave velocities during the loading and unloading cycle. The dry rock S-wave velocities exhibit a response consistent with the other samples, where the velocity slightly increases during the unloading cycle. This sample showed the most S-wave velocity response to fluid; where the brine-saturated velocity is slower than when flushed with the oil-CO_2 mixture and CO_2.

High Permeability Number 2

Much like the high-permeability number 1 sample, the high-permeability number 2 sample (Figure 8(a)) shows a very similar P-wave velocity response to live oil with a mole fraction of 0.334 CO_2 and pure CO_2. These velocities nearly converge at differential pressures of 31.03 MPa and 37.92 MPa. Also, the velocity of the sample when flushed with live oil begins to converge with the oil-CO_2 mixture and CO_2 at differential pressures of 31.03 MPa and above. Again, this is most likely due to the effect of gas coming out of solution (live oil has a bubble point estimated to be 13.33 MPa at 63.9°C).

Low Permeability Number 1

The most obvious difference between the low-permeability number 1 sample and the other core samples is the much higher dry rock velocity for both P- and S-waves when compared to the fluid-saturated and fluid-flushed velocities (Figure 9). These higher dry rock velocities are the result of being recorded at 20°C, while the fluid measurements were recorded at the estimated reservoir temperature of 63.9°C.

Low Permeability Number 2

The low-permeability number 2 sample S-wave velocities (Figure 10(b)) show a similar trend as the cemented sample where the brine-saturated velocity is at some pressures marginally higher than when the sample was flushed with live oil and CO_2. Again, typically the opposite effect is observed.

Average Axial Strain

Average axial strains measured during the dry rock loading and unloading cycle for three samples, each from a different lithological zone, are shown in Figure 11. The isostatic confining pressure has a greater effect on the high-permeability sample and least affects the cemented sample. We used these results to correct for sample length, porosity, and density at each differential pressure.

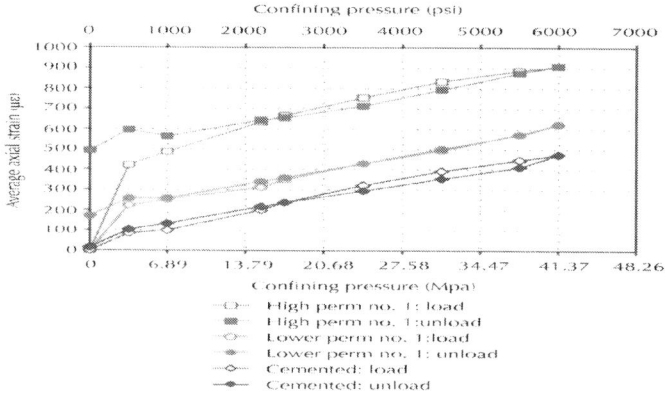

Figure 11: Average axial strain as a function of isostatic confining pressure for three dry rock core samples.

Log Facies Model

One method to link the core sample information to well log response is through log facies modeling. We generated a log facies model using 12 wells logged with the same technology within Postle Field. These well logs were corrected for environmental effects related to pressure, temperature, and borehole diameter. The well logs used as inputs for the log facies model were gamma ray, neutron porosity, bulk density, and photoelectric logs. The reservoir zones from each well log were spliced together in series to capture the variability of the reservoir. To identify the different log responses within the reservoir zone, principal component analysis was applied to maximize variability followed by cluster analysis to group data by similar log response. Figure 12 shows the box and whisker plot for each of the four clusters and the log response associated with each cluster. For each box, the central green line is the median of the data, the edges of the box are the 25th and 75th percentiles, and vertical lines (whiskers) extend to the extreme data points not considered outliers, the outliers are plotted as red diamonds. Based on the trends of the four box and whisker plots for each log response, log facies number 1 is interpreted as high-quality reservoir, log facies number 2 as intermediate-quality reservoir, log facies number 3 as low-quality reservoir, and log facies number 4 represents an interbedded shale.

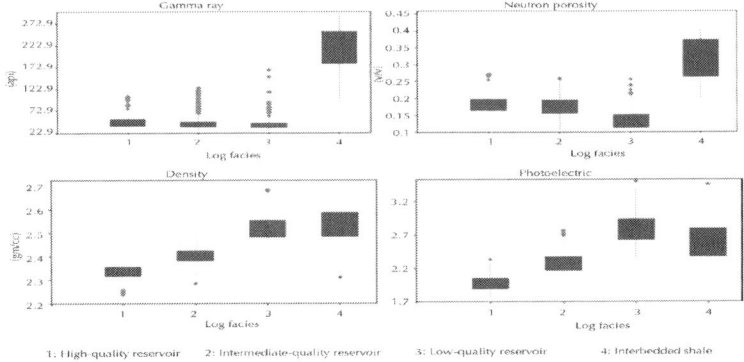

Figure 12: Box and whisker plot of the four clusters for the input logs used to generate the log facies model.

Once the log facies are established, a supervised classification method is then applied to wells with additional data such as sonic or core data. This is shown in Figure 13 where the log facies model has been applied to a well with core-measured porosity and permeability. From Figure 13, an inference can be made that the dark blue log facies (number 1) relates to high-permeability zones, the light blue log facies (number 2) correlates to intermediate-permeability zones, and the green log facies (number 3) corresponds to zones of low permeability.

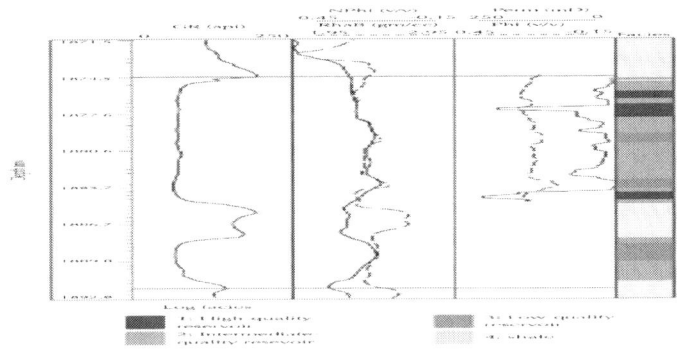

Figure 13: Log facies model applied to a well with core-measured porosity and permeability. The first track shows the gamma ray log. Track 2 plots the neutron porosity and bulk density. Track 3 represents core-measured permeability and porosity. The reservoir zone (Morrow A sandstone) is shown between the horizontal red lines.

Unfortunately, the well which the core samples used for the laboratory velocity measurements did not have the necessary input logs for supervised classification. To relate the core samples to the log facies model, we compared the permeability of the core samples to the permeability associated with each log facies taken from the well in Figure 13. Figure 14 shows this comparison where the core samples used in the laboratory velocity measurements are shown as stars and each box and whisker plot represents the variability in permeability for each of the reservoir log facies. This assumes each core sample used in the laboratory velocity measurements corresponds to a log facies. Hence, the cemented sample is associated with the low-quality reservoir log facies, the high-permeability number 1 and number 2 samples are related to the high-quality log facies and the lower-permeability number 1 and number 2 samples correspond to the intermediate-quality log facies.

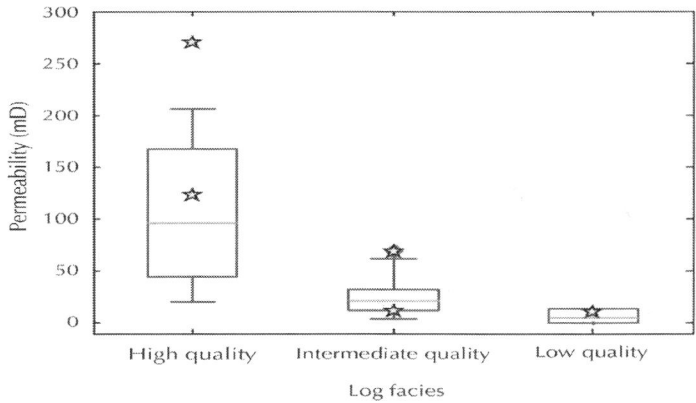

Figure 14: Permeability of the core samples used in the laboratory velocity measurements (stars) superimposed on the box and whisker plot of the per-meability related to each log facies (from the well in Figure 13).

Integrating the Log Facies Model and Core-Measured Velocity

We achieved only limited success using Gassmann's relation [9] and the effective medium theory of Mori and Tanaka [36, 37] to predict velocity changes related to fluid saturation and differential pressure in the core samples. Generally, Gassmann's relation is not expected to model fluid effects at ultrasonic frequencies. This result only means Gassmann's relation is not appropriate for this laboratory data but may be suitable at the field scale [38]. Therefore, we apply an alternative approach which assumes a direct relationship between the log facies model and the core-measured velocities. This direct relationship is based on estimating the reservoir condition in the laboratory velocity measurements, then applying the changes associated with fluid type and pressure observed in the laboratory measurements directly to the only cross-dipole velocity well logs recorded in the Postle Field study area.

The reservoir condition from petrophysical analysis estimates a water saturation of 45% [34]. Figure 15 shows the properties of live oil, brine, and the reservoir condition with a water saturation of 45%. To estimate the reservoir condition the difference in velocity between oil and brine is taken and is 76% less than the velocity of brine.

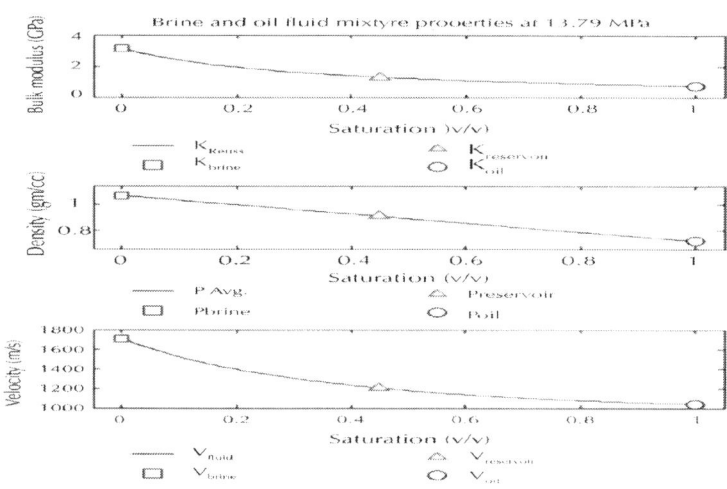

Figure 15: Calculated fluid properties at reservoir conditions for live oil, brine, and a mixture with a brine saturation of 45%.

This estimated velocity at reservoir conditions now allows for changes to be applied to the log response based on laboratory velocity measurements made on the core samples. For example, to estimate the log response at complete brine saturation, the log would be multiplied by the percent change between brine-saturated velocity and the estimated reservoir velocity from the experimental core measurement.

Since very little fluid effect is observed in the S-wave velocity core measurements, it is assumed that the fluid effect is negligible and the only changes in S-wave velocity are due to pressure. For estimating changes in the S-wave log response, the S-wave velocity is multiplied by the percent change from the reservoir condition to the desired pore pressure.

The core samples we selected to estimate the changes in P- and S-wave velocity log response were the cemented sample which is associated to the low-quality reservoir log facies, the high-permeability sample number 2 which is associated to the high-quality reservoir log facies, and the lower-permeability sample number 2 which is associated to the intermediate-quality reservoir log facies. Changes in bulk density were calculated using the density transformation equation (2) where ρ_2 is bulk density saturated with the new fluid, ρ_1 is the initial bulk density, ρ_{f1} is the density of the initial fluid, and ρ_{f2} is the density of the new fluid:

$$\rho_2 = \rho_1 + \phi(\rho_{f1} - \rho_{f2})$$
(2)

Figure 16 shows the changes in log response for the conditions of complete brine saturation at pore pressures of 6.89 MPa and 29.65 MPa. Figure 17 shows the changes in log response for the conditions of complete CO_2 saturation for pore pressures of 6.89 MPa and 29.65 MPa.

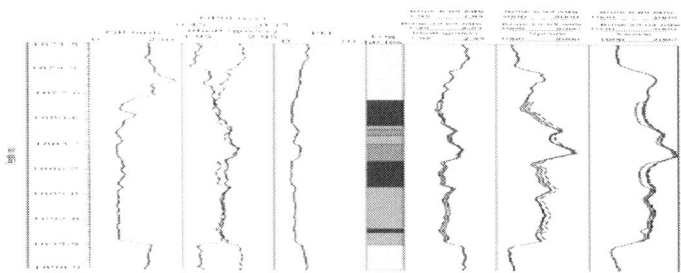

Figure 16: Estimated changes in the Morrow A sandstone for P-wave velocity, S-wave velocity, and bulk density log response for conditions of com-

plete brine saturation at pore pressures of 6.89 MPa (blue dashed lines) and 29.65 MPa (solid blue lines). The recorded in-situ P-wave velocity, S-wave velocity, and bulk density logs are shown in black.

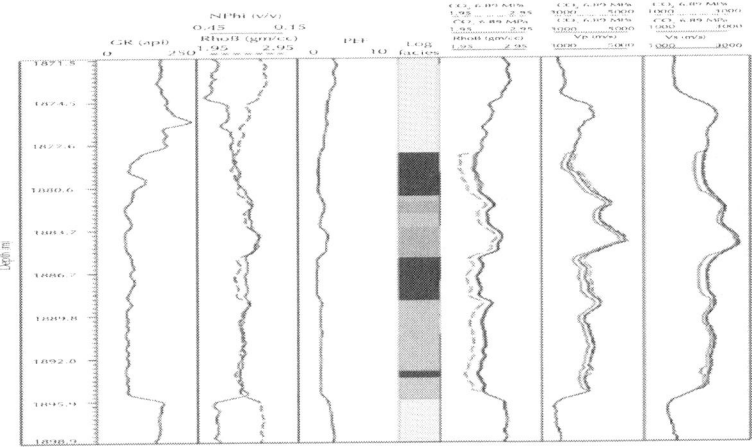

Figure 17: Estimated changes in the Morrow A sandstone for P-wave velocity, S-wave velocity, and bulk density log response for conditions of complete CO_2 saturation at pore pressures of 6.89 MPa (red dashed lines) and 29.65 MPa (solid red lines). The recorded in-situ P-wave velocity, S-wave velocity, and bulk density logs are shown in black.

Time-Lapse Amplitude Difference

We generated synthetic seismic gathers using the full waveform modeling method described in Singh and Davis [39]. To understand the P- and S-wave response at the Morrow A sandstone level, we created 1D models for the brine- and CO_2-saturated cases using the modified well logs (Figures 16 and 17). The logs were blocked at 1.5 m and simplified to have three main reflectors: the Atoka limestone above the reservoir, the Morrow A sandstone (reservoir), and a limestone layer below the reservoir. Above the Atoka limestone, the log response was smoothed to remove the effect of multiples.

From the 1D models, we generated 2D multicomponent gathers simulating vertical and horizontal sources, and receivers with vertical and horizontal components. We extracted statistical wavelets from the processed P- and S-wave seismic volumes recorded at Postle Field for use in the full waveform modeling. The P-wave wavelet was zero phase with a peak frequency of 18 Hz and the S-wave wavelet was also zero phase with a peak frequency of 13 Hz. The gathers (Figure 18) are modeled with a maximum offset of 1829 m, which is consistent with the usable offset from the seismic data recorded at Postle Field.

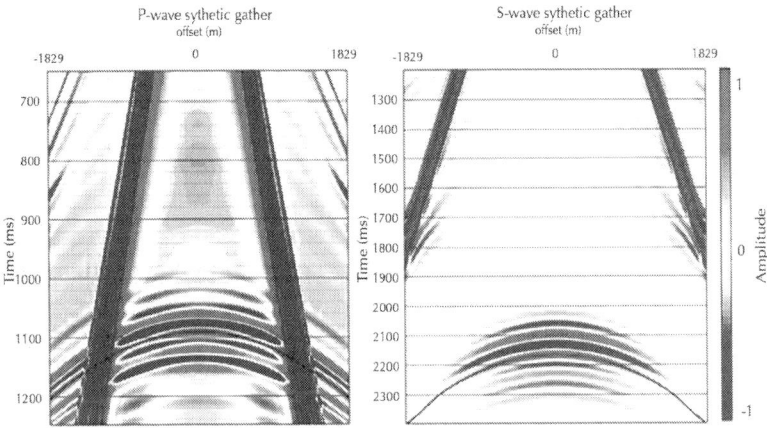

Figure 18: Full waveform synthetic gathers for P-wave (left) and S-wave (right) reflectivity. The Morrow A sandstone top is indicated by the black curve.

Seismic acquisition parameters and examples of P- and S-wave seismic data recorded at Postle Field can be found in Singh and Davis [39]. Using nearly identical wavelets and the same cross-dipole sonic logs used in this study, Singh and Davis [39] also address the tuning and interference effect at the Morrow A sandstone level.

To evaluate the changes in seismic response, we analyzed the amplitude difference of the P- and S-wave gathers for the brine- and CO_2-saturated cases. The amplitude difference for the brine-saturated case represents a pore pressure increase from 6.89 MPa to 29.65 MPa in the Morrow A sandstone. Similarly, the amplitude difference for the CO_2-saturated case represents a pore pressure increase from 6.89 MPa to 29.65 MPa in the Morrow A sandstone. The time shift for the brine and CO_2 cases was minor, calculated to be less than 0.5 milliseconds. The P-wave amplitude differences are shown in Figure 19 and the S-wave amplitude differences are shown in Figure 20.

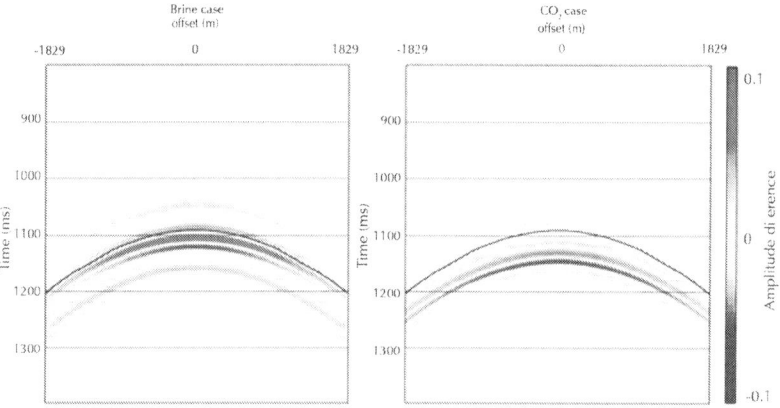

Figure 19: Comparison of the P-wave amplitude difference for the brine- and CO_2-saturated cases with a pore pressure increase from 6.89 MPa to 29.65 MPa. The Morrow A sandstone top is indicated by the black curve.

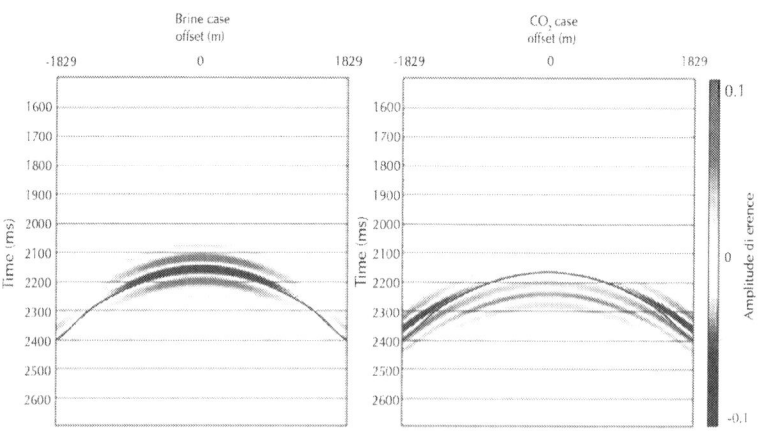

Figure 20: Comparison of the S-wave amplitude difference for the brine- and CO_2-saturated cases with a pore pressure increase from 6.89 MPa to 29.65 MPa. The Morrow A sandstone top is indicated by the black curve.

For both the brine and CO_2 cases, the P-wave modeling shows a negative amplitude difference for an increase in pressure (Figure 19), this occurs because an increase in pore pressure results in a lower acoustic impedance. For both cases, the amplitude difference is greatest at zero offset and becomes less with increasing offset. The brine-saturated case shows a large amplitude difference starting from the reflection at the Morrow A sandstone level. The CO_2 case shows a very weak amplitude decrease at the Morrow A sandstone reflection and a larger amplitude difference from the reflectors below the Morrow A sandstone. A change in waveform around the Morrow A sandstone for the CO_2 case affects the signature of the amplitude difference. These observations suggest near-angle P-wave stacks would be better suited for time-lapse analysis with larger amplitude differences occurring below the reservoir.

Increasing pore pressure lowers the elastic impedance and hence the S-wave amplitude difference is negative with increasing pore pressure for both the brine and CO_2 cases (Figure 20). The most notable observation is the amplitude difference with increasing offset between the two cases. The amplitude difference is highest at zero offset for the brine-saturated case and decreases with increasing offset. However, the CO_2-saturated case shows the amplitude difference to increase with increasing offset. Analysis on S-wave angle stacks, rather than full offset stacks, may be more effective in detecting changes in saturation and, therefore, useful for CO_2 monitoring.

The amplitude difference of the full waveform seismic modeling shows the effects fluid saturation and pressure have on P- and S-wave reflections. P-wave reflections show the near offsets to have the most amplitude difference. While S-wave reflections show a maximum amplitude difference at near offsets for the brine case and at far offsets for the CO_2 case. Also, the S-wave amplitude difference is greater than the P-wave amplitude difference for these two cases.

To determine if the time-lapse changes are mainly due to time shift or only an amplitude difference, we compared the root mean square (RMS) amplitude difference for the two cases. Since we did not take into account any time shift when we differenced the modeled synthetic seismic gathers, representing the minimum and maximum pressures for the brine and CO_2 cases (Figures 19 and 20), we calculated the RMS amplitude difference at the Morrow A sandstone top. We examined a 20 ms window centered at the Morrow a sandstone top for the P-wave gathers and a 40 ms window for the S-wave gathers. These RMS amplitude differences are shown in Figure 21 for both the brine and CO_2 cases. We observe that the RMS amplitude difference at the Morrow A top is comparable to the amplitude difference of the synthetic gathers shown in Figures 19 and 20. The RMS amplitude difference with offset also agrees with the amplitude difference of the synthetic gathers. A comparison of the RMS amplitude difference and the differenced synthetic seismic gathers demonstrate that the effect of time shift is almost negligible at the Morrow A top. The time-lapse changes at the Morrow A top (observed in Figures 19 and 20) are mainly due to amplitude differences. Figure 21 also confirms that S-wave gathers show a larger time-lapse change as compared to the P-wave gathers.

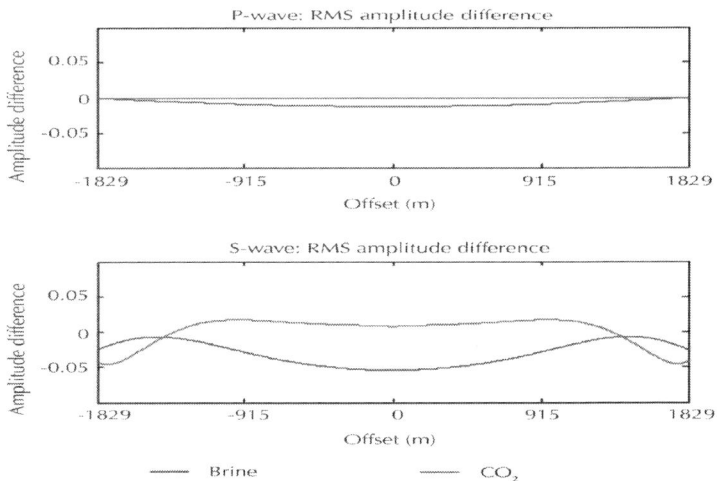

Figure 21: RMS amplitude difference at the Morrow a sandstone top for the synthetic seismic modeling of the P- and S-wave gathers.

CONCLUSIONS

Laboratory core measurements show the effects of differential pressure and fluid saturation through the P- and S-wave velocity response of five Morrow A sandstone samples. The samples were saturated with brine and flushed with live oil, an oil-CO_2 mixture, and pure CO_2. The P-wave velocity shows a sensitivity to both fluid saturation and pressure. Whilst the S-wave velocity shows a dependence to changes in differential and a nearly negligible sensitivity to fluid saturation. Core samples taken from the high-permeability zone show the most velocity change to confining pressure and pore pressure. The core sample from the cemented zone shows the least amount of velocity change to confining pressure and pore pressure. The overall velocity decrease was greater for S-waves as shown in Tables 3 and 4. Average axial strain also showed the high-permeability sample to be the most stress-sensitive and the cemented sample the least sensitive to confining pressure.

We attempted modeling the core-measured velocities using Gassmann's relation [9] and the effective medium theory of Mori and Tanaka [36, 37]. Neither of these models successfully predicted the measured P-wave or S-wave velocities. The P- and S-wave velocity well logs were modified to estimate changes due to pressure and fluid saturation based on the laboratory core-measured velocities. This approximation was implemented through a log facies model where each log facies was associated with a core sample.

The P-wave amplitude difference for the brine- and CO_2-saturated cases, modeling a pore pressure increase from 6.89 MPa to 29.65 MPa, displayed a maximum amplitude difference at the near offsets. Whereas the S-wave shows a maximum amplitude difference at the near offsets for the brine case and a maximum amplitude difference at far offsets for the CO_2 case.

The full waveform modeling results show how fluid and pressure changes in the Morrow A sandstone reservoir affect the characteristics of P- and S-wave amplitude differences. In addition, the full waveform modeling shows an advantage to using angle-limited stacks as compared to full offset stacks for time-lapse analysis. These results demonstrate the benefit and establish motivation for using S-wave seismic data in time-lapse studies involving changes in pressure and fluid saturation.

ACKNOWLEDGMENTS

The authors would like to thank the sponsors of The Reservoir Characterization Project at Colorado School of Mines, specifically Whiting Petroleum: the current operator of Postle Field; the Center for Rock Abuse at Colorado School of Mines for the use of their Rock Squeezing Laboratory, especially Professor Michael Batzle and Weiping Wang; Tom R. Bratton for his help and guidance generating the log facies model; BP America for the use of the seismic modeling code. A. V. Wondler would like to thank BP for the BP Colorado School of Mines Ph.D. Fellowship.

REFERENCES

1. L. Kootungal, "2010 worldwide EOR survey," Oil and Gas Journal, vol. 108, no. 14, pp. 41–53, 2010.

2. A. Tura and D. Lumley, "Estimating pressure and saturation changes from time-lapse AVO data," in the 69th Annual International Meeting, SEG, Expanded Abstracts, pp. 1655–1658, 1999.

3. M. Landrø, "Discrimination between pressure and fluid saturation changes from time-lapse seismic data," Geophysics, vol. 66, no. 3, pp. 836–844, 2001.

4. T. L. Davis, M. J. Terrell, R. D. Benson, R. Cardona, R. R. Kendall, and R. Winarsky, "Multicomponent seismic characterization and monitoring of the CO_2 flood at Weyburn Field, Saskatchewan," Leading Edge, vol. 22, no. 7, pp. 696–697, 2003.

5. R. J. Pawar, N. R. Warpinski, J. C. Lorenz et al., "Overview of a CO_2 sequestration field test in the West Perl Queen reservoir, New Mexico," Environmental Geosciences, vol. 13, no. 3, pp. 163–180, 2006.

6. A. Robinson and T. L. Davis, "Improving efficiency of CO_2 sequestration using short-term seismic monitoring: Applications from the Postle field EOR programme," First Break, vol. 29, no. 1, pp. 75–78, 2011.

7. A. Nur and Z. Wang, Eds., Seismic and Acoustic Velocities in Reservoir Rocks, Experimental Studies, vol. 1, chapter 1, Society of Exploration Geophysicists, 1998.

8. Z. Xue, T. Ohsumi, and H. Koide, "An experimental study on seismic monitoring of a CO_2 flooding in two sandstones," Energy, vol. 30, no. 11-12, pp. 2352–2359, 2005.

9. F. Gassmann, "Uber die elastizitat poroser medien," Vierteljahrsschrift der Naturforschenden Gesellschaft in Zrich, vol. 96, pp. 1–23, 1951.

10. G. T. Kuster and M. N. Toksoz, "Velocity and attenuation of seismic waves in two-phase media: Part I. Theoretical formulations," Geophysics, vol. 39, no. 5, pp. 587–606, 1974.

11. "Fluid Acoustics for Geophysics calculator," Rock Physics Lab, University of Houston and Center for Rock Abuse, Colorado School of Mines, 2009.

12. M. King, "Wave velocities in rocks as a function of changes in overburden pressure and pore fluid saturants," Geophysics, vol. 31, pp. 50–73, 1966.

13. S. N. Domenico, "Rock lithology and porosity determination from shear and compressional wave velocity," Geophysics, vol. 49, no. 8, pp. 1188–1195, 1984.

14. Z. Xue and T. Ohsumi, "Seismic wave monitoring of CO_2 migration in water-saturated porous sandstone," Exploration Geophysics, vol. 35, pp. 25–32, 2004.

15. Z. Xue and X. Lei, "Laboratory study of CO_2 migration in water-saturated anisotropic sandstone, based on P-wave velocity imaging," Exploration Geophysics, vol. 37, pp. 10–18, 2006.

16. J. Q. Shi, Z. Xue, and S. Durucan, "Seismic monitoring and modelling of supercritical CO_2 injection into a water-saturated sandstone: Interpretation of P-wave velocity data," International Journal of Greenhouse Gas Control, vol. 1, no. 4, pp. 473–480, 2007.

17. X. Lei and Z. Xue, "Ultrasonic velocity and attenuation during CO_2 injection into water-saturated porous sandstone: measurements using difference seismic tomography," Physics of the Earth and Planetary Interiors, vol. 176, no. 3-4, pp. 224–234, 2009.

18. J. Kim, Z. Xue, and T. Matsuoka, "Experimental study on CO_2 monitoring and saturation with combined P-wave velocity and resistivity," in International Oil and Gas Conference and Exhibition (IOGCEC '10), pp. 312–319, June 2010.

19. A. F. Siggins and D. N. Dewhurst, "Saturation, pore pressure and effective stress from sandstone acoustic properties," Geophysical Research Letters, vol. 30, no. 2, pp. 61-1–61-4, 2003.

20. A. F. Siggins, "Velocity-effective stress response of CO_2-saturated sandstones," Exploration Geophysics, vol. 37, pp. 60–66, 2006.

21. A. F. Siggins, M. Lwin, and P. Wisman, "Laboratory calibration of the seismo-acoustic response of CO_2 saturated sandstones," International Journal of Greenhouse Gas Control, vol. 4, no. 6, pp. 920–927, 2010.

22. N. I. Christensen and H. F. Wang, "The influence of pore pressure and confining pressure on dynamic elastic properties of Berea sandstone," Geophysics, vol. 50, no. 2, pp. 207–213, 1985.

23.   R. Hofmann, X. Xu, M. Batzle, M. Prasad, A. K. Furre, and A. Pillitteri, "Effective pressure or what is the effect of pressure?" Leading Edge, vol. 24, no. 12, pp. 1256–1260, 2005.

24.   X. Xu, R. Hofmann, M. Batzle, and T. Tshering, "Influence of pore pressure on velocity in low-porosity sandstone: implications for time-lapse feasibility and pore-pressure study," Geophysical Prospecting, vol. 54, no. 5, pp. 565–573, 2006.

25.   M. A. Capello, Geology and rock physics of the San Andres Formation in Vacuum Field, New Mexico, M.S. thesis, Colororado School of Mines, 1995.

26.   L. Duranti, Time-lapse multicomponent seismic analysis of reservoir dynamics, Ph.D. dissertation, Colorado School of Mines, 2001.

27.   L. T. Brown, Integration of rock physics and reservoir simulation for the interpretation of time-lapse seismic data at Weyburn Field, Saskatchewan, M.S. thesis, Colorado School of Mines, 2002.

28.   H. Yamamoto, Using time-lapse seismic measurments to improve ow modeling of CO_2 injection in the Weyburn Field: a naturally fractured, layered reservoir, Ph.D. dissertation, Colorado School of Mines, 2004.

29.   E. Rojas, Elastic rock properties of tight gas sandtones for reservoir characterization at Rulison Field, Colorado, M.S. thesis, Colorado School of Mines, 2005.

30.   Z. Wang and A. M. Nur, "Effects of CO_2 flooding on wave velocities in rocks with hydrocarbons," SPE Reservoir Engineering, vol. 4, no. 4, pp. 429–436, 1989.

31.   S. Sonnenberg, "Tectonic and sedimentation model for Morrow sandstone deposition, Sorrento Field, Denver Basin, Colorado," Mountain Geologist, vol. 22, pp. 180–191, 1985.

32.   J. Benton, "Subsurface stratigraphic analysis, Morrow (Pennsylvanian), North Central Texas County, Oklahoma," The Shale Shaker Digest VII, vol. 21–23, pp. 1–28, 1973.

33.   T. D. Jobe, Optimizing geo-cellular reservoir modeling in a braided river incised valley fill: Postle Field, Texas County, Oklahoma, M.S. thesis, Colorado School of Mines, 2010.

34.   A. E. Heris, Integrated ow simulation and time-lapse seismic reservoir characterization in an enhanced oil recovery project,

Postle Field, Texas County, Oklahoma, Ph.D. dissertation, Colorado School of Mines, 2011.

35. W. Wang, M. Batzle, and A. Wandler, "Effects of CO_2 on brine, gas, and oil properties," in 2009 Annual Meeting of Fluid/DHI, Rock Physics Lab, University of Houston and Center for Rock Abuse, Colorado School of Mines, 2009.

36. T. Mori and K. Tanaka, "Average stress in matrix and average elastic energy of materials with misfitting inclusions," Acta Metallurgica, vol. 21, no. 5, pp. 571–574, 1973.

37. Y. Benveniste, "A new approach to the application of Mori-Tanaka's theory in composite materials,"Mechanics of Materials, vol. 6, no. 2, pp. 147–157, 1987.

38. G. Mavko, "Expert answers," CSEG Recorder, vol. 30, no. 5, pp. 8–12, 2005.

39. P. Singh and T. Davis, "Advantages of shear wave seismic in Morrow sandstone detection,"International Journal of Geophysics, vol. 2011, Article ID 958483, 16 pages, 2011.

Is There Deep-Seated Subsidence in the Houston-Galveston Area?

Jiangbo Yu, Guoquan Wang, Timothy J. Kearns, and
Linqiang Yang

Department of Earth and Atmospheric Sciences, National Center for Airborne LiDAR Mapping, 312 Science & Research Building 1, Room 312, University of Houston, Houston, TX 77204-5007, USA

ABSTRACT

Long-term continuous groundwater level and land subsidence monitoring in the Houston-Galveston area indicates that, during the past two decades (1993–2012), the groundwater head has been increasing and the overall land subsidence rate has been decreasing. Assuming that the hydraulic head in the aquifer will reach or exceed the preconsolidation level in the near future, will subsidence in the

Houston-Galveston area eventually cease? The key to answer this question is to identify if there is deep-seated subsidence in this area. This study investigated the recent subsidence observed at different depths in the Houston-Galveston area. The subsidence was recorded by using 13 borehole extensometers and 76 GPS antennas. Four of the GPS antennas are mounted on the deep-anchored inner pipes of borehole extensometers. We conclude that recent subsidence (1993–2012) in the Houston-Galveston area was dominated by the compaction of sediments within 600 m below the land surface. Depending on the location of specific sites, the compaction occurred within the Chicot aquifer and part or all of the Evangeline aquifer. No measurable compaction was observed within the Jasper aquifer or within deeper strata. Deep-seated subsidence is not likely occurring in the Houston-Galveston area.

INTRODUCTION

The Houston-Galveston area provides one of the most extreme case studies of subsidence hazards, which is a problem that affects many other U.S. metropolitan areas, for example, Los Angeles (CA), Sacramento (CA), New Orleans (LA), Phoenix (AZ), and Tucson (AZ) [1–3]. The Houston Ship Channel and the Galveston Bay area have experienced severe subsidence [4, 5]. Up to four meters of subsidence has occurred in the area along the Houston Ship Channel from 1915–17 to 2001 [5]. The area along the Houston Ship Channel includes Pasadena, Bay Town, Seabrook, and Texas City. Land subsidence has increased the frequency and severity of infrastructure damage and flooding in the Houston-Galveston area. It has been well recognized that recent subsidence in the Houston-Galveston area has occurred as a direct result of ground water withdrawals [6]. Historical subsidence caused by oil and gas production in the Houston-Galveston area was also studied by researchers [7]. Different from ground water, the depth from which the oil and gas are produced would affect the amount of possible subsidence. It is because the deeper the clays are, the more compacted they are and the less they can compact following a reduction in liquid or gas pressure. Holzer and Bluntzer [8] studied subsidence profiles across 29 oil and gas fields in Houston, Texas, from 1943 to 1973. They concluded that the contribution of petroleum withdrawal to local land subsidence is relatively small compared to aquifer compaction. Since 1980s, significant oil and gas production has moved off-shore in the Gulf of Mexico area [9]. Most on-shore oil reservoirs are now in their tertiary recovery phase and producers need to exploit enhanced oil recovery (EOR) techniques [10] to maintain production. By reinjecting CO_2 and saline water into the reservoir, the loss of reservoir pressure is minimized and therefore the subsidence is limited. However, the ground water extraction remains an issue in the Houston-Galveston area. The ground water has been withdrawn for the purpose of meeting the needs of municipal supply, industry, and irrigation. Heavy groundwater pumping depressurizes and dewaters the major aquifers (Chicot and Evangeline) in this area, which causes compaction of the clay sediment layers of the aquifers [11, 12]. A study conducted by the U.S. Geological Survey (USGS) reported that, from 1943 to 1977, the groundwater withdrawals in the Houston-Galveston area resulted in a decline of water levels of as much as 76 m in wells completed in the Chicot aquifer and as much as 91 m in wells completed in the Evangeline aquifer [4].

In order to prevent land subsidence, the Texas Legislature created the Harris-Galveston Subsidence District (HGSD) in 1975, "...for the purpose of ending subsidence, which contributes to or precipitates flooding, inundation, or overflow of the district, including without limitation rising waters resulting from storms or hurricanes" [13]. HGSD was authorized to issue (or refuse) well permits, promote water conservation and education, and promote conversion from ground-water to surface-water supplies. The Texas Legislature created the Fort Bend Subsidence District (FBSD) in 1989 [14] and the Lone Star Groundwater Conservation District (LSGCD) in 2001 [15] to manage groundwater resources and minimize land subsidence occurring in the Fort Bend County and the Montgomery County, respectively. These three districts (HGSD, FBSD, and LSGCD) operate a campaign style GPS network, which includes over 70 GPS stations. These GPS stations are called Port-A-Measure (PAM) stations. The earliest PAM sites were installed in 1993 [16]. The Texas Department of Transportation and other local agencies operate over 20 continuously operating reference stations (CORS) in the Houston-Galveston area [17]. The USGS has been operating 13 borehole extensometers at 11 sites since 1973 for the purpose of observing compaction of aquifers in this area [18].

From measurements recorded by 13 extensometers and 72 GPS stations in the Houston-Galveston area, we derived the average subsidence rate (mm/year) in this area from 2006 to 2012 (Figure 1). The locations of these observational sites are plotted in Figure 1. The red lines represent the locations of about 150 principle faults mapped by USGS in the Houston area using airborne LIDAR data collected after tropical storm Allison in 2001 [19]. The tan blobs represent the locations of salt domes [20]. All GPS stations used for this study have a history of three years or longer.

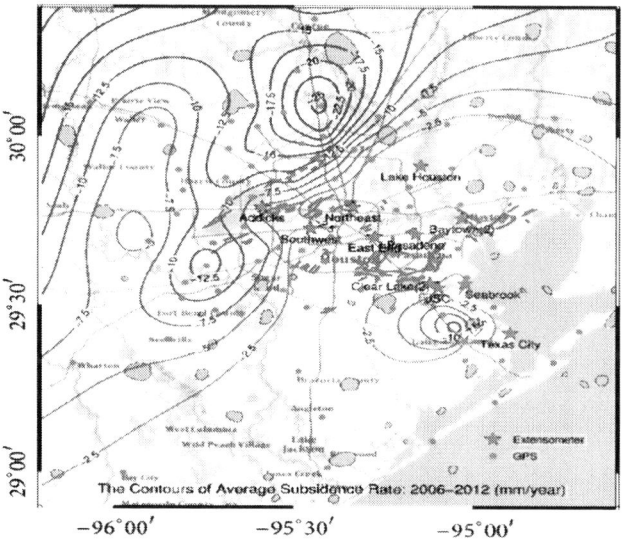

Figure 1: Map showing contours of the average subsidence rate (mm/year) during the time span from 2006 to 2012. The contours are derived from subsidence observed from 72 GPS stations and 13 borehole extensometers. The red lines represent the locations of about 150 principle faults mapped by USGS in the Houston area using airborne LIDAR data collected after tropical storm Allison in 2001 [19]. The tan blobs represent the locations of salt domes [20].

The subsidence in Galveston, downtown Houston, and the area along the Houston Ship Channel has almost stabilized (<3 mm/year). Subsidence greater than 10 mm/year and up to 25 mm/year is still occurring in land areas north and west of Houston, which include Spring, Jersey Village, Addicks, Katy, and the Sugar Land area. These areas are located in Regulatory Area 3 of the HGSD's 1999 and 2013 regulation plans [13, 21] or Area A of the 2003 FBSD regulation plan [14]. There was no groundwater regulation in these areas until 2010.

There are three primary aquifers in the Houston-Galveston area. They are the Chicot, Evangeline, and Jasper aquifers, which all belong to the Gulf Coast aquifer. There is a confining layer (Burkeville) between the Evangeline and Jasper aquifers (Figure 2). The youngest and uppermost portion of the Chicot aquifer consists of Holocene and Pleistocene age sediments; the underlying Evangeline aquifer consists of Pliocene and Miocene age sediments [22, 23]. The Chicot and Evangeline aquifers are composed of laterally discontinuous deposits of gravel, sand, silt, and clay. The oldest and most deeply buried Jasper aquifer consists of Miocene age sediments. The two shallower aquifer units, the Chicot and Evangeline, have been described as being hydrologically connected [22]. This means that changes in the hydraulic properties of one aquifer will affect the properties of the other. However, the Jasper aquifer is not hydrologically connected to the Evangeline aquifer. This is because the Burkeville confining layer greatly restricts the vertical flow of groundwater from one aquifer to the other.

(a)

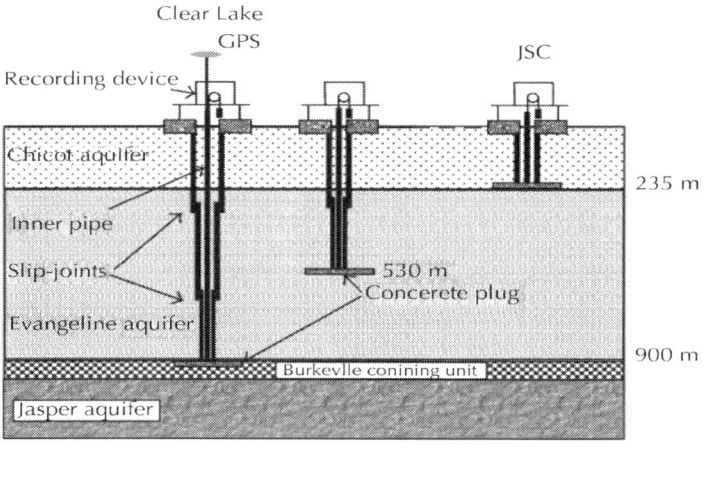

(b)

Figure 2: (a) A site photo showing the deep borehole extensometer and the collocated GPS station (TXEX) at the Clear Lake site. The GPS antenna is mounted on the extended inner pipe of the extensometer borehole, which is anchored in the strata 936 m below the ground surface. (b) A sketch depicting the local aquifers and borehole extensometers (−936 and −530 m) at Clear Lake and a nearby shallow extensometer (−235 m) at Johnson Space Center (JSC). The horizontal distance between the two extensometers at the Clear Lake site is 100 m. The horizontal distance between the Clear Lake and JSC sites is 2.5 km.

The borehole extensometers in the Houston-Galveston area were designed as "double pipe wells." These wells were drilled to preselected depths. Figure 2 illustrates a diagram depicting a cross-sectional perspective of the borehole extensometers at the Clear Lake and the Johnson Space Center (JSC) sites. The compaction of the interval between the land surface and the bottom of the inner pipe is continuously monitored by an analog recorder. Thus, extensometers are also called "compaction recorders." There is an opening screen (about 3 m in height) at the bottom of the outer casing of each borehole extensometer. This design allows each extensometer borehole to also function as a groundwater monitoring well, called a piezometer. The scientific theory and operation of borehole extensometers are explained by Poland and Yamamoto [24] and Gabrysch [25].

Figure 3 illustrates the close correlation between the compaction of aquifers and the change of groundwater level recorded at 13 extensometer sites operated by USGS (1973–2012). Red lines represent boreholes that were completed in the Chicot aquifer. Blue lines represent the boreholes that were completed in the Evangeline aquifer. Surface water from Lake Livingston became available to the Houston Ship Channel area and the coastal area of Galveston in late 1976 [4]. This water was used to augment existing groundwater withdrawals. After 1976, groundwater withdrawals decreased significantly in both Galveston County and southeastern Harris County. The groundwater levels at all extensometer sites, except at Addicks, have been rising during the past four decades. As a result, the rates of aquifer compaction at the eastern and southeastern sites have been decreasing since 1990, about 14 years after starting to reduce groundwater withdrawals, as indicated by the shaded area in Figure 3(a). At the Texas City and Pasadena City sites the land surface has even rebounded slightly since 1990. The groundwater level of the Evangeline aquifer at the Addicks site had been continuously decreasing (2.5 m/year) until 2001. As a result, the compaction at the Addicks site occurred continuously from 1973 to 2012. However, the compaction rate has decreased since 2001. The groundwater level is currently rising at the Addicks site at an average rate of 3 m/year. The recent (2001–2012) subsidence rate at the Addicks site has declined to less than 1 cm/year from previous 4 cm/year (1974–2000) (Figure 4(a)). Compaction at this site will continuously occur as long as the groundwater head remains below the preconsolidation level (a critical value below which inelastic compaction of the fine-grained deposits occurs) within the aquifer systems in this area. The aquifer systems in this area include both clay and silt layers.

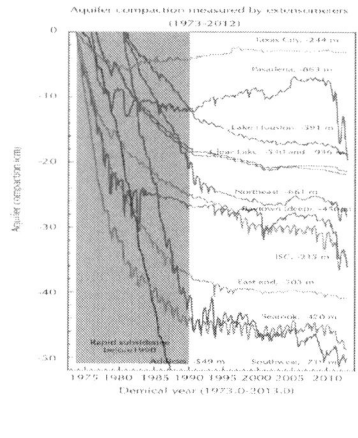

(a)

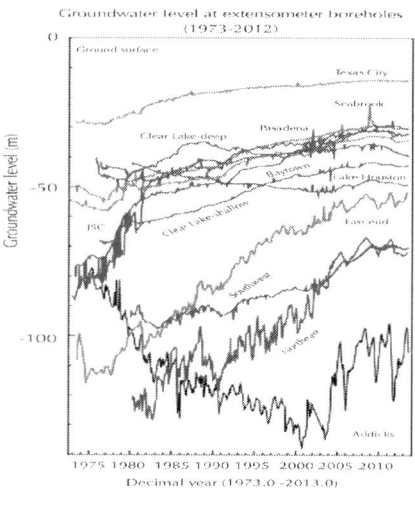

(b)

Figure 3: Plots depicting the history of aquifer compaction (a) and the level of groundwater head (b) observed at 13 extensometer sites in the Houston-Galveston area. The depth of each borehole is marked in the left figure. Red lines represent boreholes that were completed within the Chicot aquifer. Blue lines represent boreholes that were completed within the Evangeline aquifer.

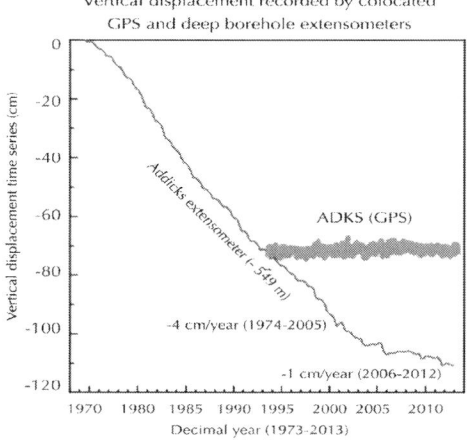

(a)

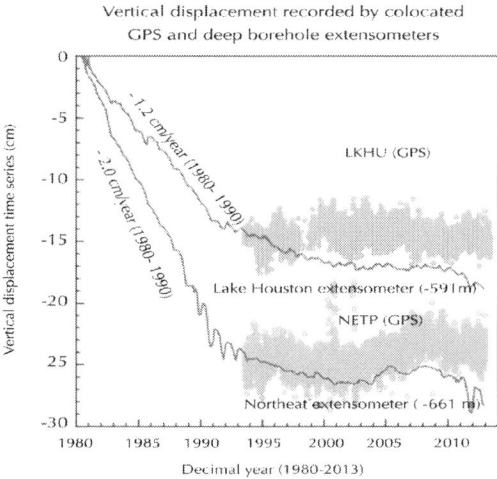

(b)

Figure 4: Comparisons of aquifer compaction recorded by extensometers and vertical displacements recorded by three GPS antennas mounted on the inner pipes of extensometer boreholes.

Figures 1 and 3 indicate that the groundwater regulations enforced by the subsidence districts have largely succeeded in their primary objective of limiting subsidence in the Houston-Galveston area. Long-term groundwater monitoring conducted by the USGS has indicated that the overall groundwater levels in both the Chicot and Evangeline aquifers have been increasing during the past two decades [18]. The districts' ability to limit subsidence raises a question: will the subsidence in the Houston-Galveston area eventually stop if the groundwater level reaches or exceeds the preconsolidation level in this area? This is an important question to consider for the long-term management of land and groundwater resources in the Houston-Galveston area. The same question has been asked in New Orleans. Researchers have found that there is deep-seated (or tectonic-controlled) subsidence caused by deep-rooted faulting in the New Orleans area in addition to shallow subsidence associated with aquifer compaction [26–28]. Consequently, the subsidence in the New Orleans area is not likely to cease in the near future. This will be a great challenge to long-term coastal land management. The goal of this study is to explore if deep-seated subsidence is occurring in the Houston-Galveston area.

OBSERVATIONS FROM COLOCATED GPS AND EXTENSOMETERS

In 1993, the HGSD installed GPS antennas on the extended inner pipes of three borehole extensometers for the purpose of providing stable reference stations in the Houston-Galveston area. The three extensometers are located at Addicks, Lake Houston, and Northeast (Figure 1). There are two closely spaced (100 m) borehole extensometers at Clear Lake. In 2010, a GPS antenna was installed on the extended inner pipe of the deeper borehole extensometer at the Clear Lake site (Figure 2). The four GPS units located at Addicks, Lake Houston, Northeast, and Clear Lake are named ADKS, LKHU, NETP, and TXEX, respectively. The bottom of the inner pipe of each extensometer borehole was firmly anchored to the stratum at the bottom of the borehole. The depth of the boreholes at Addicks (ADKS), Lake Houston (LKHU), Northeast (NETP), and Clear Lake (TXEX) sites are 549, 591, 661, and 936 m below the land surface, respectively. The Addicks borehole was completed at the top of the Burkeville confining layer [4]. According to the sediment profile of the Houston-Galveston area provided by USGS [23], the extensometer boreholes at the Lake Houston and Northeast sites also reach the bottom of the Evangeline aquifer. In order to exclude the effect of superficial soil deformation associated with moisture, temperature, and biological changes, the concrete platform on the land surface around the borehole was built on piers bored six meters below the land surface [16, 29]. As a result of this design, the extensometer measures aquifer compaction between the bottom of the borehole and six meters below the land surface. The GPS antenna reference point (ARP) measures the displacement as it translates to the anchored point at the end of the borehole inner pipe. Thus, the GPS antenna is able to measure the vertical displacement below the bottom of the borehole, which is called deep-seated subsidence in this paper. If such deep-seated subsidence does occur, it could be a combined result of the compaction of the underlying aquifer (Jasper), the consolidation of sediments below the Jasper aquifer, and possibly vertical displacement associated with tectonic movements. The GPS daily measurements are obtained by using the GIPSY/OASIS (V6.2) software package developed at the Jet Propulsion Laboratory. It uses precise point positioning (PPP) with single receiver phase ambiguity resolutions [30, 31]. According to our previous investigations, the daily PPP solutions would achieve 2–4 mm horizontal accuracy and 6–8 mm vertical accuracy in the Houston area [17, 32]. The displacement time series of the vertical component represents the positional changes of the GPS antenna relative to the center of mass of the Earth.

Figure 4 illustrates the compaction (1975, 1980–2012) recorded by the extensometers and the deep-seated subsidence (1993–2013) recorded by the GPS antennas at the Addicks, Lake Houston, and Northeast sites. The extensometer-recorded compaction times series indicates that there has been 40 cm compaction of the shallow aquifers at the Addicks site and minor compaction (<5 cm) at the Lake Houston and Northeast sites since 1993. However, the GPS displacement time series indicates that there has been no considerable vertical displacement (subsidence) below the Evangeline aquifers at all three sites since 1993. Hence, the compaction recorded by extensometers at Addicks, Lake Houston, and Northeast sites accounts for the total subsidence at each site.

OBSERVATIONS FROM CLOSELY SPACED EXTENSOMETERS

With the closely spaced extensometers completed at different depths, we are able to estimate the variation of compaction rate in depth. Figure 5(a) illustrates the long-term (1975–2012) aquifer compaction recorded by two very closely spaced (100 m) extensometers at Baytown, which were completed at different depths (−131 and −450 m). According to the combined hydrogeological information published by USGS [22, 33], the depths of the bottom of the Chicot and Evangeline aquifers in this area are about 250 and 800 m, respectively. Accordingly, the shallow borehole at the Baytown site was ended in the middle of the Chicot aquifer and the deep borehole was ended within the top one-third of the Evangeline aquifer. The compaction time series indicates that there was rapid compaction during the first 10 years of the monitoring period (1973 to 1983). The compaction accumulated within the Chicot aquifer (−6 m to −131 m, 15 mm/year) accounts for 50% of the total compaction (−6 m to −450 m, 30 mm/year) during this time period. The total compaction includes the compaction within the whole Chicot aquifer and one-third of the Evangeline aquifer. Since there is no GPS antenna installed at the top of the inner pipe of the deep borehole, it is difficult to determine if the compaction recorded by the deep extensometer represents the total subsidence at this site. The compaction at the Baytown site was insignificant (1.5–2 mm/year) at both shallow and deep borehole sites during the period from 1984 to 2009. However, it appears that rapid aquifer compaction has been occurring since 2010. The rapid compaction rates recorded by the shallow and deep extensometers are 50 mm/year and 20 mm/year, respectively. The two extensometer boreholes are only 100 m apart and the site-specific deformation within the top six meters has been excluded from extensometer recording. Thus, it is expected that the compaction recorded by the shallow extensometer should be no larger than the compaction recorded by the deep extensometer. The anomaly of compaction curves at the Baytown site since 2010 implies that one or both extensometers have not been working properly since 2010.

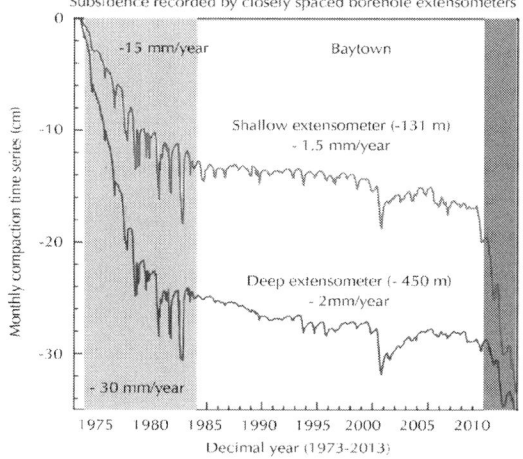

(a)

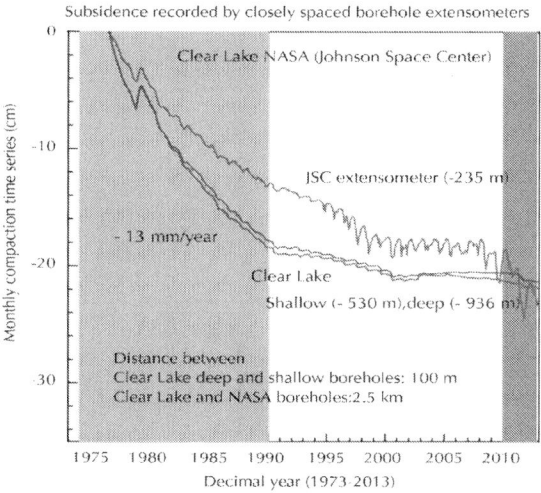

(b)

Figure 5: Aquifer compaction (1973–2013) recorded by five extensometers at Baytown, Clear Lake, and the Johnson Space Center (JSC).

Figure 5(b) illustrates the compaction time series recorded by two closely spaced extensometers at Clear Lake and one extensometer at Johnson Space Center. A vertical profile about the depths of the three borehole extensometers and the local aquifers is illustrated in Figure 2. The two borehole extensometers at Clear Lake were completed in the Evangeline aquifer at 530 m and 936 m below the land surface, respectively. The horizontal distance between the two extensometers is about 100 m. The extensometer borehole at the Johnson Space Center (JSC) was completed 235 m below the land surface, which reached the bottom of the Chicot aquifer. The deeper borehole at Clear Lake was completed in the Burkeville confining layer (−900 m) [25]. The two extensometers (−530 m and −936 m) at the Clear Lake site consistently recorded the same amount of compaction from 1976 to 2012. This indicates that there was no aquifer compaction between 530 m and 936 m. In other words, the compaction at the Clear Lake site occurred within sediments shallower than 530 m below the land surface, which includes the whole Chicot aquifer and the top part of the Evangeline aquifer. The JSC extensometer is about 2.5 km southeast of the Clear Lake site. The cumulative compaction recorded at the JSC extensometer site reached the compaction recorded at the Clear Lake sites by the end of 2011. The observations at the Baytown, Clear Lake, and JSC sites suggest that most of the compaction was occurring in the Chicot aquifer and the top portion of the Evangeline aquifer.

TECTONIC SUBSIDENCE

The Houston metropolitan area, and more broadly the Gulf Coast in general, has numerous gravitationally induced normal faults. Active faults in Houston were first recognized in as early as 1926 [7]. By 1973, about 52 faults had been recognized in the Houston area [34–36]. These faults had an aggregate length of 220 km [37]. By 1979, the identified historically active faults had increased to over 150 in number and to more than 500 km in aggregate length [38]. Numerous subsurface faults have been documented beneath the Houston metropolitan area at depths from 800 m to 1250 m [38]. Some of these subsurface faults have affected shallower sediments and offset the present land surface. Unfortunately, active faults in the Houston area have not been thoroughly investigated for decades. This is because USGS budgets for mapping the faults in this area were eliminated in the late 1970s. As a result, most of the published research on active faulting dates from that time. Structural damage on or near the Earth's surface associated with ground deformation, such as damage to buildings, bridges, roads, and pipelines, has been frequently reported every year in the Houston area. The frequent superficial damage may imply that some faults are active. However, there are few quantitative studies about the depth of the fault creeping in the affected area.

Data from GPS sites in the Houston-Galveston area have been analyzed in this study to investigate the possible tectonic movement. Figure 6 illustrates long-term three-component positional time series derived from the GPS stations: ADKS, LKHU, NETP, and TXEX. The horizontal components, north-south (NS) and east-west (EW), have been transformed to the stable Houston reference frame (SHRF) from the original IGS08 reference frame [32]. The SHRF is defined by nine long-history continuously operating reference stations (CORS) outside the Houston area. The root-mean-squares (RMS) of the positional time series from all stations are less than one centimeter. We calculated the RMS value using data from 1993–2005 for stations ADKS, LKHU, and NETP. The RMS value is also called repeatability or accuracy of GPS measurements in the geodesy community [39, 40]. The three-component positional time series of these four GPS antennas indicate the antennas have been stable (total displacement <2 cm, except NETP-NS) in both vertical and horizontal directions during the past 20 years (1993–2012). This indicates that the strata at the bottom (about −600 m) of the boreholes did not experience any considerable subsidence.

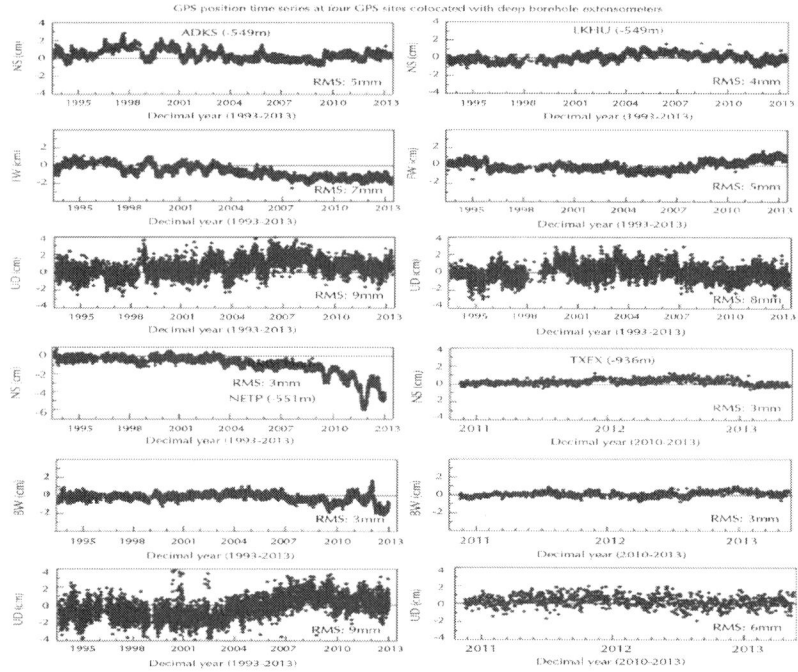

Figure 6: Three-component displacement time series recorded by four GPS antennas mounted on the inner pipes of borehole extensometers at Addicks (ADKS), Lake Houston (LKHU), Northeast (NETP), and Clear Lake (TXEX).

Figure 7 illustrates the average horizontal and vertical ground displacement vectors derived from recent GPS observations (2006–2012) in the Houston-Galveston area. The horizontal ground movement vectors have been transformed to the SHRF. The vertical ground displacement vectors are relative to the mass center of the Earth. The distribution of displacement vectors indicates that both horizontal and vertical ground displacements are site specific. The overall spatial correlations between the horizontal and vertical ground displacements are currently weak. There are no consistent horizontal local movements. A detailed review of positional time series from all GPS time series indicates that very few sites show steady horizontal movements. The overall horizontal movements are minor (<5 mm/year). The maximum subsidence rate is about 25 mm/year. It should be noted that potential superficial ground deformations within the top 6 m have been excluded from GPS observations since most GPS antennas are mounted on antenna poles that are anchored about 6 m below the land surface. Certain near fault ground displacements may have been missed from GPS observations because most GPS stations are at least a few kilometers away from the fault traces.

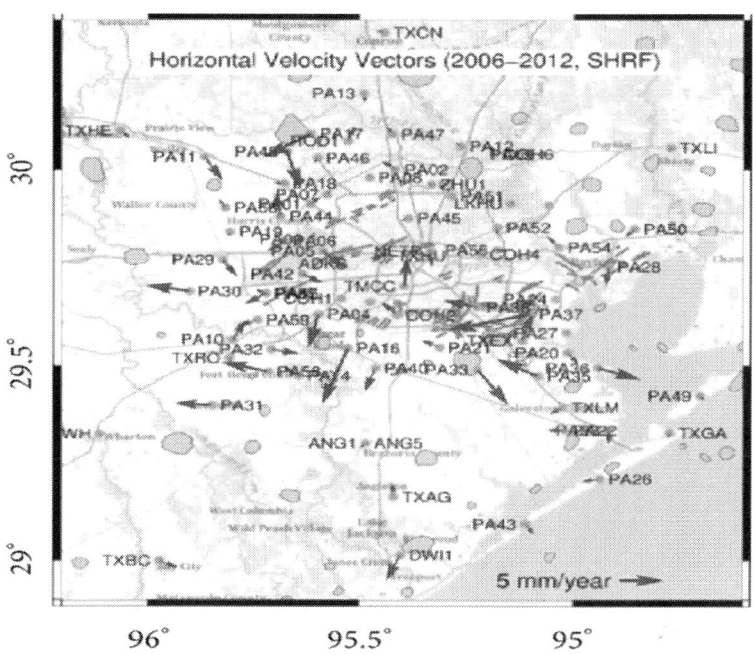

(a)

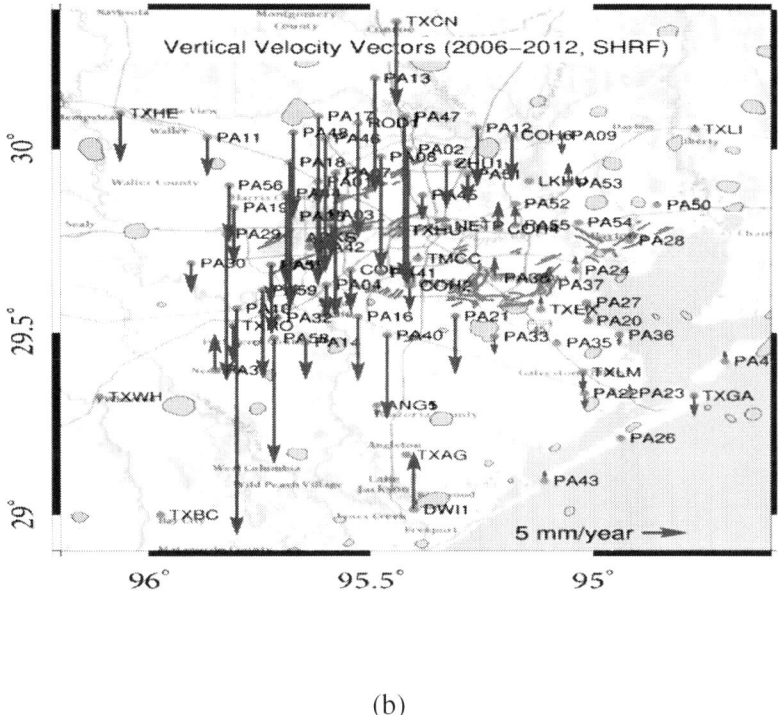

(b)

Figure 7: Recent horizontal and vertical ground deformation velocity vectors in the Houston-Galveston area derived from extensometer and GPS observations (2006–2012). The horizontal velocity vectors have been transformed to the stable Houston reference frame (SHRF). The red lines represent the locations of about 150 principle faults mapped by USGS in the Houston area using airborne LIDAR data collected after tropical storm Allison in 2001 [19]. The tan blobs represent the locations of salt domes [20].

Dokka et al. [27, 28] proposed that deep-rooted faulting (tectonic) activities are dominating the subsidence in the New Orleans area. Several other studies [41–44] argued the "rapid" tectonic subsidence rates presented in Dokka's works. However, many researchers agree that the New Orleans has been subsiding for a long period and the subsidence is not dominated by human activities. Understanding the occurrence of the tectonic subsidence is critical for designing an appropriate flood protection strategy in the New Orleans coastal area. Figure 8 illustrates vertical displacement time series (2002–2012) recorded at three permanent GPS stations along the New Orleans coastal area and four GPS stations along the Galveston coastal area. The three stations in New Orleans show steady subsidence with a rate of 4 mm/year. Dokka et al. [28] proposed that the steady subsidence is associated with a 7–10 km thick active allochthon that is detached from the stable North America plate. New Orleans lies on top of this active allochthon. The four stations on the Galveston coast do not show any considerable subsidence during the same time span (2002–2012). The GPS observations presented in Figure8 indicate that the steady subsidence currently occurring in New Orleans is not occurring in Galveston.

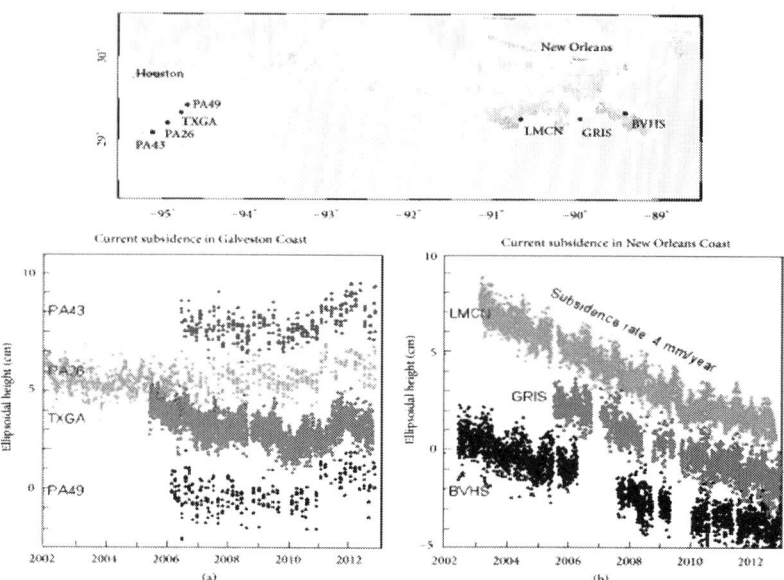

Figure 8: Comparisons of land subsidence currently occurring on the coast of Galveston, TX, (a) and on the coast of New Orleans, LA (b).

CONCLUSION AND DISCUSSION

This investigation indicates that the recent subsidence (1993–2012) in the Houston-Galveston area is dominated by the compaction of aquifers associated with groundwater withdrawals. The affected aquifers are limited to about 600 m below the land surface, which would include the Chicot and part or all of the Evangeline aquifer depending on the site where subsidence was monitored. The Chicot and Evangeline aquifers are the primary sources of the municipal groundwater supply in the Houston-Galveston area [3]. No measurable compaction was recorded in the Jasper aquifer or its underlying sediments. Deep-seated subsidence is not likely occurring in the Houston-Galveston area.

This study also indicates that the present subsidence rate could vary considerably even within a small area of a few square kilometers. This could be caused by differences in groundwater withdrawal near each site or by different clay-to-sand ratios in the subsurface sediments. Therefore, it is difficult to precisely extrapolate or infer a rate of subsidence for adjacent areas on the basis of the subsidence rate measured at a specific GPS or extensometer site. Hence, the contours of subsidence illustrated in Figure 1 should be interpreted with caution.

ACKNOWLEDGMENTS

The authors thank USGS and HGSD for providing extensometer and GPS data to the public. The authors appreciate many thoughtful discussions with Mr. Mark G. Kasmarek at USGS. This study was supported by an NSF CAREER Award EAR-1229278 and an NSF MRI Award EAR-1242383.

REFERENCES

1.   D. Galloway, D. Jones, and S. Ingebritsen, "Land subsidence in the United States," U.S. Geological Survey, U.S. Department of the Interior, Reston, Va, USA, 1999.

2.   R. Engelkemeir, S. D. Khan, and K. Burke, "Surface deformation in Houston, Texas using GPS,"Tectonophysics, vol. 490, no. 1-2, pp. 47–54, 2010. · ·

3.   G. Bawden, M. Johnson, and M. Kasmarek, Investigation of land subsidence in the Houston-Galveston region of Texas by using the Global Positioning System and Interferometric Synthetic Aperture, 1993–2000, U.S. Geological Survey, Reston, Va, USA, 2012.

4. R. K. Gabrysch, Ground-Water Withdrawals and Land-Surface Subsidence in the Houston-Galveston Region, Texas, 1906-80, Texas Department of Water Resources, Austin, Tex, USA, 1984.

5. M. C. Kasmarek, R. K. Gabrysch, and M. R. Johnson, Estimated Land-Surface Subsidence in Harris County, Texas, 1915–17 to 2001, U.S. Geological Survey, Reston, Va, USA, 1915.

6. R. K. Gabrysch and C. W. Bonnet, Land-Surface Subsidence in the Houston-Galveston Region, Texas, Texas Water Development Board, Austin, Tex, USA, 1975.

7. W. Pratt and D. Johnson, "Local subsidence of the goose creek oil field," The Journal of Geology, vol. 34, pp. 577–590, 1926.

8. T. L. Holzer and R. L. Bluntzer, "Land subsidence near oil and gas fields, Houston, Texas.," Ground Water, vol. 22, no. 4, pp. 450–459, 1984. · ·

9. C. Cleveland, "A brief history of offshore oil drilling," 2013,http://www.eoearth.org/view/article/160618/.

10. A. T. F. S. Gaspar Ravagnani, E. L. Ligero, and S. B. Suslick, "CO_2 sequestration through enhanced oil recovery in a mature oil field," Journal of Petroleum Science and Engineering, vol. 65, no. 3-4, pp. 129–138, 2009. · ·

11. M. C. Kasmarek, M. R. Johnson, and J. K. Ramage, Water-Level Altitudes 2010 and Water-Level Changes in the Chicot, Evangeline, and Jasper Aquifers and Compaction 1973–2009 in the Chicot and Evangeline Aquifers, Houston–Galveston Region, Texas, U.S. Geological Survey, Reston, Va, USA, 2010.

12. M. R. Johnson, J. K. Ramage, and M. C. Kasmarek, Water-Level Altitudes 2011 and Water-Level Changes in the Chicot, Evangeline, and Jasper Aquifers and Compaction 1973–2010 in the Chicot and Evangeline Aquifers, Houston-Galveston Region, Texas, U.S. Geological Survey, Reston, Va, USA, 2011.

13. Harris-Galveston Subsidence District, Harris-Galveston Subsidence District Regulatory Plan, Adopted April 14, 1999, Amended September 12, 2001 and June 9, 2010, Harris-Galveston Subsidence District, Friendswood, Tex, USA, 1999.

14. Fort Bend Subsidence District, Fort Bend Subsidence District 2003 Regulation Plan, Fort Bend Subsidence District, Richmond, Tex, USA, 2003.

15. Lone Star Groundwater Conservation District, Long Start Groundwater Conservation District Management Plan, Long Start Groundwater Conservation District, Conroe, Tex, USA, 2008.

16. D. B. Zilkoski, L. W. Hall, G. J. Mitchell, et al., "The Harris-Galveston Coastal Subsidence District/National Geodetic Survey automated global positioning system subsidence monitoring Project," in Proceedings of the U.S. Geological Survey Subsidence Interest Group Conference, pp. 13–28, U.S. Geological Survey, Galveston, Tex, USA, 2003.

17. G. Wang and T. Soler, "Using opus for measuring vertical displacements in Houston, Texas," Journal of Surveying Engineering, vol. 139, no. 3, pp. 126–134, 2013. · ·

18. M. C. Kasmarek, M. R. Johnson, and J. K. Ramage, Water-Level Altitudes 2013 and Water-Level Changes in the Chicot, Evangeline, and Jasper Aquifers and Compaction 1973–2012 in the Chicot and Evangeline Aquifers, Houston-Galveston Region, Texas, U.S. Geological Survey, Reston, Va, USA, 2013.

19. S. Shah and J. Lanning-Rush, Principal Faults in the Houston, Texas, Metropolitan Area, U.S. Geological Survey, Reston, Va, USA, 2005.

20. American Association of Petroleum Geologists, Salt Tectonism of the U.S. Gulf Coast Basin, Tulsa, Okla, USA, 2011.

21. Harris-Galveston Subsidence District, Harris-Galveston Subsidence District Regulatory Plan 2013, Harris-Galveston Subsidence District, Friendswood, Tex, USA, 2013.

22. E. T. Baker, Stratigraphic and Hydrogeologic Framework of Part of the Coastal Plain of Texas, Texas Department of Water Resources, Austin, Tex, USA, 1979.

23. E. T. Baker, Hydrology of the Jasper Aquifer in the Southeast Texas Coastal Plain, Texas Water Development Board, Austin, Tex, USA, 1986.

24. J. F. Poland and S. Yamamoto, "Field measurement of deformation," in Guidebook to Studies of Land Subsidence due to Ground-Water Withdrawal, J. F. Poland, Ed., pp. 17–35, United Nations Educational Scientific and Cultural Organization, Paris, France, 1984.

25. R. K. Gabrysch, "Subsidence in the Houston-Galveston region, Texas, USA," in Guideb. to Stud. L. Subsid. due to Ground-Water Withdrawal, J. F. Poland, Ed., pp. 253–262, United Nations Educational Scientific and Cultural Organization, Paris, France, 1984.

26. R. K. Dokka, "Modern-day tectonic subsidence in coastal Louisiana," Geology, vol. 34, no. 4, pp. 281–284, 2006. · ·

27. R. K. Dokka, "The role of deep processes in late 20th century subsidence of New Orleans and coastal areas of southern Louisiana and Mississippi," Journal of Geophysical Research B: Solid Earth, vol. 116, no. 6, Article ID B06403, 2011. · ·

28. R. K. Dokka, G. F. Sella, and T. H. Dixon, "Tectonic control of subsidence and southward displacement of southeast Louisiana with respect to stable North America," Geophysical Research Letters, vol. 33, no. 23, Article ID L23308, 2006.

29. G. Wang, J. Yu, T. J. Kearns, and J. Ortega, "Assessing the accuracy of long-term subsidence derived from borehole extensometer data using GPS observations: a case study in Houston, Texas," Journal of Surveying Engineering, 2014. ·

30. J. F. Zumberge, M. B. Heflin, D. C. Jefferson, M. M. Watkins, and F. H. Webb, "Precise point positioning for the efficient and robust analysis of GPS data from large networks," Journal of Geophysical Research B: Solid Earth, vol. 102, no. 3, Article ID 96JB03860, pp. 5005–5017, 1997. · ·

31. W. Bertiger, S. D. Desai, B. Haines et al., "Single receiver phase ambiguity resolution with GPS data,"Journal of Geodesy, vol. 84, no. 5, pp. 327–337, 2010. · ·

32. G. Wang, J. Yu, J. Ortega, G. Saenz, T. Burrough, and R. Neill, "A stable reference frame for the study of ground deformation in the Houston metropolitan area, Texas," Journal of Geodetic Science, vol. 3, pp. 13–27, 2013.

33. M. C. Kasmarek, M. R. Johnson, and J. K. Ramage, Water-level altitudes 2012 and water-level changes in the Chicot, Evangeline, and Jasper aquifers and compaction 1973–2011 in the Chicot and Evangeline aquifers, Houston–Galveston region, Texas, U.S. Geological Survey, Reston, Va, USA, 2012.

34. P. Weaver and M. M. Sheets, "Active faults, subsidence, and foundation problems in the Houston, Texas, area," in Geology of the Gulf Coast and Central Texas, and Guidebook of Excursions, E. H. Rainwater and R. P. Zingula, Eds., pp. 254–265, Houston Geological Society, Houston, Tex, USA, 1962.

35. D. C. van Siclen, "The Houston fault problem," in Proceedings of the 3rd annual Convention, Texas Section, pp. 9–31, American Institute of Professional Geologists, Houston, Tex, USA, 1967.

36. M. M. Sheets, "Active surface faulting in the Houston area, Texas," Houston Geological Society Bulletin, vol. 13, pp. 24–33, 1971.

37. W. M. Reid, Active faults in Houston, Texas [Ph.D. thesis], University of Texas at Austin, 1973.

38. E. R. Verbeek, K. W. Ratzlaff, and U. S. Clanton, Faults in Parts of North-Central and Western Houston Metropolitan Area, Texas, Texas, U.S. Geological Survey, Reston, Va, USA, 1979.

39. G. Wang, "GPS landslide monitoring: single base vs. network solutions—a case study based on the Puerto Rico and Virgin Islands permanent GPS network," Journal of Geodetic Science, vol. 1, no. 3, pp. 191–203, 2011. ·

40. G. Wang and T. Soler, "Opus for horizontal subcentimeter-accuracy landslide monitoring: case study in the Puerto Rico and Virgin Islands Region," Journal of Surveying Engineering, vol. 138, no. 3, pp. 143–153, 2012. · ·

41. K. T. Milliken, J. B. Anderson, and A. B. Rodriguez, "A new composite Holocene sea-level curve for the northern Gulf of Mexico," in Response of Upper Gulf Coast Estuaries to Holocene Climate Change and Sea-Level Rise, J. B. Anderson and A. B. Rodriguez, Eds., pp. 1–12, Geological Society of America Special Papers, Boulder, Colo, USA, 2008.

42. T. E. Törnqvist, D. J. Wallace, J. E. A. Storms et al., "Mississippi Delta subsidence primarily caused by compaction of Holocene strata," Nature Geoscience, vol. 1, no. 3, pp. 173–176, 2008. · ·

43. A. S. Kolker, M. A. Allison, and S. Hameed, "An evaluation of subsidence rates and sea-level variability in the northern Gulf of Mexico," Geophysical Research Letters, vol. 38, no. 21, Article ID L21404, 2011.

44. A. R. Simms, J. B. Anderson, R. DeWitt, K. Lambeck, and A. Purcell, "Quantifying rates of coastal subsidence since the last interglacial and the role of sediment loading," Global and Planetary Change, vol. 11, pp. 296–308, 2013.

Citations

CHAPTER 1

Albina Mukhametshina and Elena Martynova, "Electromagnetic Heating of Heavy Oil and Bitumen: A Review of Experimental Studies and Field Applications," Journal of Petroleum Engineering, vol. 2013, Article ID 476519, 7 pages, 2013. doi:10.1155/2013/476519.

CHAPTER 2

Biji Shibulal, Saif N. Al-Bahry, Yahya M. Al-Wahaibi, Abdulkader E. Elshafie, Ali S. Al-Bemani, and Sanket J. Joshi, "Microbial Enhanced Heavy Oil Recovery by the Aid of Inhabitant Spore-Forming Bacteria: An Insight Review," The Scientific World Journal, vol. 2014, Article ID 309159, 12 pages, 2014. doi:10.1155/2014/309159.

CHAPTER 3

Abhishek Punase, Amy Zou, and Riza Elputranto, "How Do Thermal Recovery Methods Affect Wettability Alteration?,"Journal of Petroleum Engineering, vol. 2014, Article ID 538021, 9 pages, 2014. doi:10.1155/2014/538021.

CHAPTER 4

Dingwei Zhu, Jichao Zhang, Yugui Han, Hongyan Wang, and Yujun Feng, "Laboratory Study on the Potential EOR Use of HPAM/VES Hybrid in High-Temperature and High-Salinity Oil Reservoirs," Journal of Chemistry, vol. 2013, Article ID 927519, 8 pages, 2013. doi:10.1155/2013/927519.

CHAPTER 5

Saurabh Mishra, Achinta Bera, and Ajay Mandal, "Effect of Polymer Adsorption on Permeability Reduction in Enhanced Oil Recovery," Journal of Petroleum Engineering, vol. 2014, Article ID 395857, 9 pages, 2014, doi:10.1155/2014/395857.

CHAPTER 6

Yong Tang, Ruizhi Yang, and Xiaoqiang Bian, "A Review of CO_2 Sequestration Projects and Application in China," The Scientific World Journal, vol. 2014, Article ID 381854, 11 pages, 2014. doi:10.1155/2014/381854.

CHAPTER 7

Aaron V. Wandler, Thomas L. Davis, and Paritosh K. Singh, "An Experimental and Modeling Study on the Response to Varying Pore Pressure and Reservoir Fluids in the Morrow A Sandstone," International Journal of Geophysics, vol. 2012, Article ID 726408, 17 pages, 2012. doi:10.1155/2012/726408.

CHAPTER 8

Jiangbo Yu, Guoquan Wang, Timothy J. Kearns, and Linqiang Yang, "Is There Deep-Seated Subsidence in the Houston-Galveston Area?" International Journal of Geophysics, vol. 2014, Article ID 942834, 11 pages, 2014. doi:10.1155/2014/942834.

Index

C

Calcium carbonate 56, 57, 60, 63, 64, 69
Calcium soap 62
Carbon Capture 123, 126, 144, 145, 151
Carbon capture and sequestration (CCS) 124
Carbon sequestration with enhanced oil recovery (CSEOR) 129
Central Drug House (CDH) 102
Chemical enhanced oil recovery (CEOR) 25
Chemically enhanced oil recovery (CEOR) 80

Chemical property 129
Coal bed methane (CBM) 138
Continuously operating reference stations (CORS) 189, 201
Crude oil 21
Cyclic steam stimulation (CSS) 65, 68

D

Department of Energy (DOE) 27

E

Electromagnetic energy 8, 9, 10
Electromagnetic heating 2, 3, 4, 5, 7, 9, 10, 11, 12, 15, 16, 17, 18, 19
Electromagnetic radiation 2, 12

Energy structure 124, 126
Enhanced oil recovery (EOR) 23, 154, 189
Enhanced Solvent Extraction Incorporating Electromagnetic Heating technology (ESEIEH) 15
Erucyl dimethyl amidobetaine (EDAB) 80

F

Fluid Acoustics for Geophysics (FLAG) 164

G

Gas oil ratio (GOR) 164
Geological media 127, 136, 143
Global warming 140

H

Harris-Galveston Subsidence District (HGSD) 189
Heavy reliance 123, 143
Hydrolyzed polyacrylamide (HPAM) 80

L

Lone Star Groundwater Conservation District (LSGCD) 189

M

Microbial enhanced oil recovery (MEOR) 22, 42

O

Oil fraction 58
Oil production 22, 60, 68, 69, 71, 127, 128, 129, 132, 133, 134
Oil saturation 55, 56, 63, 65, 66, 68, 71
Original oil in place (OOIP) 67, 69, 115

P

Partially hydrolyzed polyacrylamide (PHPA) 99, 121
Partially hydrolyzed Polyacrylamide (PHPA) 102
Partially Hydrolyzed Polyacrylamide (PHPA) 101
Petroleum Development Oman (PDO) 35
Physical property 64

R

Reservoir Characterization Project (RCP) 155
Residual resistance factor (RRF) 109
Resistance factor (RF) 109
Result show 118
Rock surface 54, 57, 58, 59, 61, 62, 63, 68
Root mean square (RMS) 179

S

Small number 111
Sodium bicarbonate 69
Soil organic carbon 140
Solid surface 54, 62, 63
Stable Houston reference frame (SHRF) 201, 204
Stock tank oil initially in place (STOIIP) 66

T

Thermally assisted gas oil gravity drainage (TAGOGD) 36
Thermal recovery 54, 57, 65, 66, 67, 73, 75, 76

V

Viscoelastic surfactant (VES) 82
Viscosity 1, 10

W

Water alternating gas (WAG) 159
Water production 100, 121